Conceptual Physics Experiments

Editor in Chief

Lv Hongfeng

CHINA AGRICULTURAL UNIVERSITY PRESS

内容简介

全书共分为4章。第一章是绪论。第二章阐述测量误差和数据处理的基础知识。第三章为基础性实验，主要包括与理论课课堂教学密切相关的实验。第四章为综合性及创新性实验，由实验目的、实验原理、实验仪器、实验内容及步骤、数据处理、实验思考题几个部分组成。

图书在版编目(CIP)数据

大学物理实验 = Conceptual Physics Experiments：英文/吕洪凤主编. —北京：中国农业大学出版社，2013.9

ISBN 978-7-5655-0787-8

Ⅰ.①大… Ⅱ.①吕… Ⅲ.①物理学-实验-双语教学-高等学校-教材-英文 Ⅳ.①O4-33

中国版本图书馆 CIP 数据核字(2013)第 175604 号

书　名	Conceptual Physics Experiments
作　者	Lv Hongfeng　主编
策划编辑　宋俊果　潘晓丽	责任编辑　洪重光
封面设计　郑　川	责任校对　王晓凤　陈　莹
出版发行　中国农业大学出版社	
社　　址　北京市海淀区圆明园西路2号	邮政编码　100193
电　　话　发行部 010-62818525,8625	读者服务部 010-62732336
编辑部 010-62732617,2618	出 版 部 010-62733440
网　　址　http://www.cau.edu.cn/caup	e-mail　cbsszs @ cau.edu.cn
经　　销　新华书店	
印　　刷　涿州市星河印刷有限公司	
版　　次　2013年9月第1版　2013年9月第1次印刷	
规　　格　787×980　16开本　15.5印张　286千字	
定　　价　30.00元	

图书如有质量问题本社发行部负责调换

Editor in Chief:
    Lv Hongfeng

Associate Editor:
    Liu Yuying

Contributors:
    Sang Hongyi
    Wang Xiaosu

Editorial Committee:
    Huang Guanhua
    Xu Tingwu
    Feng Weizhe
    Jiao Qunying
    Wang Ning

前　言

"大学物理实验"通常是本科生进入大学后的第一门实验课,主要是培养学生基本实验技能、延伸课堂理论教学,该课程在培养高素质创新人才、创建高水平大学方面起着其他课程不可替代的重要作用。

为了培养高质量、高素质的国际化复合型人才,物理实验课程(Conceptual Physics Experiments)全英文授课从2009年起开始试讲。为保证学生们的国际化学习环境,所有的物理实验讲义和报告均用英文撰写。经过几年的试运行,我们深深体会到:新生在物理实验方面的基本英语词汇比较缺乏、对物理专业术语不熟悉,学生在听课中遇到很大困难,在实验过程中动手能力差,缺乏独立思考和解决问题的能力。

针对试行教学中出现的问题,并积极借鉴国内外先进教材的编写经验,我们编写了本教材。全书共分为4章。第一章是绪论。第二章阐述了测量误差和数据处理的基础知识。第三章为基础性实验,主要包括与理论课堂教学密切相关的实验。第四章为综合性及创新性实验,各实验由实验目的、实验原理、实验仪器、实验内容及步骤、数据处理、实验思考题几个部分组成。

本书将为同学们提供和现有实验设计紧密结合的指导教材,同时也提供了一个学习物理英语的机会。该教材有针对性地启发学生,引导学生逐步分析问题并通过科学的实验方法最终解决问题,从而逐步提高分析物理现象的能力以及理论联系实际的工作能力。教材全面覆盖了力、波、光、电、磁等基础物理知识,以便学生从看起来零散的各个实验中,系统地领悟实验的思想、方法、技术和应用,拓宽学生的知识面,扩充实验教学的信息量。

本教材的主要特色与创新体现在如下几个方面:

1. 受众群体明确。针对试行教学中出现的问题,学习国内外先进教材的经验,为各专业全英语授课的同学提供全英文的物理实验课教材—《Conceptual Physics Experiments》,也可供非全英文的双语教学的本科生做参考。

2. 启发性强。结合仪器的实物照片,介绍各个实验的仪器原理图、结构、运行机理、使用方法等;在实验仪器使用与维护部分总结各个实验运行中常见故障,分析其原因和排除方法;为启发学生思考,提供多个思考题,以完成由单一学科学习型向培养综合应用型人才转变。

3. 实用与创新并重。编写内容将涵盖常见的物理实验,创新型实验的内容将着重体现新知识与技术。

4. 形式新颖。本书采取简练、直观、图文并茂、通俗易懂的编写形式,着重于该课程在各专业的实际应用。

5. 配套齐全。制作与课程配套的多媒体课件和实验演示动画并实现网上在线学习和演示(http://211.82.81.110/wlsy)。

参编人员多讲授过"Conceptual Physics Experiments"及全英文"Physics"理论课程,积累了较丰富的英语教学经验。电学(及刚体、霍尔效应)、光学、磁学部分分别由刘玉颖、桑红毅、王小素编写,其他内容由吕洪凤编写。

编者

2013 年 6 月

Contents

1 INTRODUCTION ········ (1)
2 DATA AND ERROR ········ (3)
  2.1 Measurement and Measurement Errors ········ (3)
  2.2 Characteristics of Errors ········ (5)
    2.2.1 Indeterminate Errors/Random Errors ········ (5)
    2.2.2 Determinate Errors/Systematic Errors ········ (15)
  2.3 Measurement Uncertainty ········ (21)
  2.4 Significant Figures ········ (33)
  2.5 Methods of Data Analysis ········ (39)
    2.5.1 Graphing ········ (39)
    2.5.2 Linear Relations ········ (47)
    2.5.3 Non-linear relations ········ (53)
3 CLASSICAL EXPERIMENTS ········ (55)
  3.1 Mechanics ········ (55)
    3.1.1 Measurement: Length and Density of a Solid Substance ········ (55)
    3.1.2 Collisions: Conservation of Linear Momentum ········ (67)
    3.1.3 Measurement of Rotational Inertia of Rigid Body ········ (75)
    3.1.4 The Determination of Liquid Surface Tension Coefficient ········ (83)
    3.1.5 Measuring Young's Modulus with a Stretching Method ········ (87)
    3.1.6 Pendulum Measurements ········ (99)
  3.2 Thermal and Wave ········ (107)
    3.2.1 Interference Coefficient of Thermal Expansion Measurement ········ (107)
    3.2.2 Determining the Specific Heat Capacity of Air ········ (111)
    3.2.3 Standing Waves ········ (117)
    3.2.4 Acoustic Resonance ········ (123)

3.3 Optics .... (129)

    3.3.1 Measurement of the Wave Length of Light .... (129)

    3.3.2 Single and Double-slit Interference .... (135)

    3.3.3 Polarization of Light .... (147)

    3.3.4 Sugar Identification Using Polarimetry .... (153)

3.4 Electricity .... (157)

    3.4.1 Measurement of Electromotive Force with Compensation Method .... (157)

    3.4.2 Direct Current, Alternating Current Measurement and Ohm's Law .... (163)

    3.4.3 The Features and the Use of the Digital Oscilloscope .... (171)

3.5 Magnetism .... (181)

    3.5.1 Magnetic Force Due to a Current-carrying Wire .... (181)

    3.5.2 Magnetic Field Measured by Hall Effect .... (189)

    3.5.3 Transformer .... (195)

    3.5.4 Faraday's Law of Induction .... (201)

4 INNOVATION EXPERIMENTS .... (207)

4.1 Optical Tweezers .... (207)

4.2 Scanning Tunneling Microscope (STM) .... (219)

4.3 Measuring Planck Constant .... (231)

APPENDIX .... (237)

REFERENCES .... (239)

REFERENCE WEBSITES .... (240)

1 INTRODUCTION

1. General Information

"The truth is, the science of Nature has been too long made only a work of the brain and the fancy. It is now high time that it should return to the plainness and soundness of observations on material and obvious things."——R. Hooke

2. Laboratory Objectives

The laboratory work associated with Physics course has two principal goals: To give you hands-on experience with the phenomena and models you will study in class; To develop basic experimental and analytic skills that will be used throughout your career in the future. The experiments can strengthen the students' hands-on ability and inspire their innovative consciousness. They provide an emphasis and point of view which physics course lacked.

The laboratory exercises that you will do here are not "experiments", in the sense of forays into the unknown designed and executed by an intrepid person. Rather, they were chosen to illustrate physical phenomena, ingenious techniques or useful methods. They were not intended to be extremely precise, and your results will be far from exact. You will be evaluated on your understanding of the material and your approach to problems, not merely the precision of your results, and you should allocate your effort accordingly.

It will emphasize very basic skills. You should develop the ability to carry out common laboratory procedures correctly and safely; To make measurements and report your results in physically meaningful form, including estimates of uncertainties where appropriate; To recognize when equipment or procedures are not working, and undertake logical corrective action. You will also have the opportunity to communicate your results in the form of short reports on each experiment.

1 INTRODUCTION

1. General Information

"The truth is, the science of Nature has been too long made only a work of the brain and the fancy. It is now high time that it should return to the plainness and soundness of observations on material and obvious things." —R. Hooke.

2. Laboratory Objectives

The laboratory work associated with Physics course has two principal goals.: To give you hands-on experience with the phenomena and models you will study in class. To develop basic experimental and analytic skills that will be used throughout your career in the future. The experiments can strengthen the students' hands-on ability and inspire their innovative consciousness. They provide an emphasis and point of view which physics cannot lacked.

The laboratory exercises that you will do here are not "experiments", in the sense of forays into the unknown designed and executed by an intrepid person. Rather, they were chosen to illustrate physical phenomena, ingenious techniques or useful methods. They were not intended to be extremely precise, and your results will be far from exact. You will be evaluated on your understanding of the material and your approach to problems, not merely the precision of your results, and you should allocate your effort accordingly.

It will emphasize yet basic skills. You should develop the ability to carry out common laboratory procedures correctly and safely; To make measurements and report your results in physically meaningful form, including estimate of uncertainties where appropriate; To recognize when equipment or procedures are not working, and undertake logical corrective action. You will also have the opportunity to communicate your results in the form of short report on each experiment.

2 DATA AND ERROR

2.1 Measurement and Measurement Errors

With every measurement, no matter how carefully it is made, there is an associated error inherent with the measurement; no one can ever exactly measure the true value of a quantity. The magnitude of the error is due to the precision of the measuring device, the proper calibration of the device, and the competent application of the device. This is different than a gross mistake or blunder. A blunder is due to an improper application of the measuring device, such as a misreading of the measurement. Careful and deliberate laboratory practices should eliminate most blunders.

Precision and Accuracy

To determine the error associated with a measurement, scientists often refer to the precision and accuracy of the measurement. Let us use the analogy of the marksman who uses a gun to fire bullets at a target. In this analogy, the gun is the instrument, the marksman is the operator of the instrument, and the results are determined by the location of the bullet holes in the target.

The precision of an experiment is a measure of the reliability of the experiment, or how reproducible the experiment is. In figure 1(a), we see that the marksman's instrument was quite precise, since his results were uniform due to the use of a sighting scope. However, the instrument did not provide accurate results since the shots were not centered on the target's bull's eye. The fact that his results were precise, but not accurate, could be due to a misaligned sighting scope, or a consistent operator error. Therefore precision tells us something about the quality of the instrument's operation.

The accuracy of an experiment is a measure of how closely the experimental results agree with a true or accepted value. In figure 1(b), we see a different experimental result. Here, the shots are centered on the bull's eye but the results were not uniform, indicating that the marksman's instrument displayed good accuracy but poor precision. This could be the result of a poorly manufactured gun barrel. In this case, the marksman will never achieve both accuracy and precision, even if he very carefully uses the instrument. If he is not satisfied with the results he must change his equipment. There-

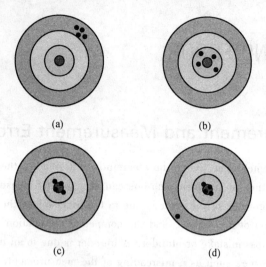

Figure 1 Precision and accuracy of an experiment
From: http://www.lepla.org

fore accuracy tells us something about the quality or correctness of the result.

We desire our results to be both precise and accurate. As shown in figure 1(c), the shots are all uniform and centered on the bull's eye. This differs from the first figure in that the marksman has compensated for the poorly aligned sighting scope.

One benefit of taking many measurements of a single property is that blunders are easily detected. In figure 1(d) we see that the results are both accurate and precise with the exception of an obvious blunder. Because several measurements were made, we can discount the errant data point as an obvious mistake, probably due to operator error.

Experimental errors are of two types: ①indeterminate (or random) and ②determinate (or systematic) errors. A measurement with relatively small indeterminate error is said to have high **precision.** A measurement with small indeterminate error and small determinate error is said to have high **accuracy.** Precision does not necessarily imply accuracy. A precise measurement may be inaccurate if it has a determinate error.

2.2 Characteristics of Errors

2.2.1 Indeterminate Errors/Random Errors

When repeated measurements of a quantity do not yield the same value, there may be some erratic influences on the measurement, or in the measuring process, larger than the smallest readable unit on the scale. The repeated measurements show a scatter of values. The scatter may have limited extent, so the measurement isn't completely uncertain. But we cannot predict (determine) exactly what the next measured value will be. Therefore these uncertainties (errors) are called **indeterminate**. The influences which affect the data so as to make values consistently too large or too small are also be called **random errors**. Among many possible causes of indeterminate errors are:

(1) Attempting to read an instrument scale to a very high precision. For example, trying to read a scale to the nearest 1/10 of its smallest division requires difficult estimation which may be highly unreliable.
(2) Mechanical irregularities in the measuring instrument. For example, readings taken from a beam balance may be affected by friction and wear of its mechanical parts. Such effects may be reduced but never completely eliminated.
(3) Uncontrolled outside influences on the apparatus.
(4) Careless technique or observation by the experimenter.

Whatever the cause, indeterminate errors reveal themselves when repeated measurements give different values. A typical set of values for a measurement might look like this:

$$3.69 \quad 3.68 \quad 3.67 \quad 3.69 \quad 3.68 \quad 3.69 \quad 3.66 \quad 3.67$$

We assume that the measured quantity has just one precise value, independent of the measuring process, and that the variability of the recorded values is caused by imperfections of the instruments or procedure. We want to represent our knowledge, obtained from these measurements, as one "best" value.

We might represent this measurement by the mean (average) of the measured values. The mean is 3.67875 exactly. Recognizing that all of these digits are not meaningful, we round this off to 3.68, retaining the certain digits 3.6 and the first uncertain one, the 8.

The value 3.68 is still somewhat uncertain. Nothing to the right of the 8 was certain enough to keep, but the 8 itself is a borderline case; it is not completely meaningless. We could be more precise and say that our value 3.68 is not likely to be "wrong" by more than ±0.02. The value ±0.02 is an estimate of the uncertainty in the value 3.68. Such results are written in standard form: 3.68 ± 0.02.

The first number is the experimenter's best estimate of the true value. The last number is a measure of the "maximum error".

The true value: One can think of the "true" value in two equivalent ways:

(1) The true value is the value one would measure if all sources of error were completely absent.

(2) The true value is the average of an infinite number of repeated measurements.

Since uncertainties can never be completely eliminated, and we have no time to take an infinite set of measurements, so we never obtain "true" values. With better equipment and more care we can narrow the range of scatter, and therefore improve our estimates of average values, confident that we are approaching "true" values more and more closely. More importantly, we can make useful estimates of how close the values are likely to be to the true value.

Error: The number following the symbol is the experimenter's estimate of how far the quoted value might deviate from the "true" value.

Estimates or measures of uncertainty are called errors. In this case we have quoted a "maximum error" in the value. We will introduce other, better, kinds of error measures later.

Error analysis is an essential part of the experimental process. It allows us to make meaningful quantitative estimates of the reliability of results. A laboratory investigation done without concern for error analysis cannot properly be called a scientific experiment.

Two other words, "deviation" and "discrepancy" are also used in science with very specific meanings:

Deviation: Suppose a set of measurements has been averaged. The magnitude of the difference between a particular measurement and the average is called the deviation of that measurement from the average. Deviations may be expressed as percents.

Example 1: The deviation of the first value in the set discussed above is $3.69 - 3.68 = 0.01$. The percent deviation is $100\% \times 0.01/3.68 = 0.2\%$.

Discrepancy: The discrepancy between any two measurements is the magnitude of their difference.

Discrepancies may also be expressed as percents. The term "discrepancy" is usually used when several independent experimental determinations of the same quantity are compared. When a student compares a lab measurement or result with the value given in the textbook, the difference is called the "experimental discrepancy". Never make the mistake of calling this comparison an "error". Textbook or handbook values are not "true" values; they are merely someone else's experimental values. If several experimenters quote values of the same quantity, and have estimated their errors properly, the discrepancy between their values should be no larger than their estimated maximum errors.

Example 2: Suppose that in the preceding example, the "textbook value" of the quantity was 3.675. Our best average experimental value was 3.68. The discrepancy is therefore 0.005. Note that the maximum error is still 0.02. Do not confuse an error with a discrepancy.

Indeterminate errors are present in all experimental measurements. The name "indeterminate" indicates that there's no way to determine the size or sign of the error in any individual measurement. Indeterminate errors cause a measuring process to give different values when that measurement is repeated many times (assuming all other conditions are held constant to the best of the experimenter's ability). Indeterminate errors can have many causes, including operator errors or biases, fluctuating experimental conditions, varying environmental conditions and inherent variability of measuring instruments.

The effect that indeterminate errors have on results can be somewhat reduced by taking repeated measurements then calculating their average. The average is generally considered to be a "better" representation of the "true value" than any single measurement, because errors of positive and negative sign tend to compensate each other in the averaging process.

Propagation of Indeterminate Errors

Indeterminate errors have unknown sign. If their distribution is symmetric about the mean, then they are unbiased with respect to sign. Also, if indeterminate errors in different quantities are independent of each other, their signs have a tendency offset each other in computations.

When we are only concerned with **limits of error** (or maximum error) we must assume a "worst-case" combination of signs. In the case of subtraction, $A - B$, the worst-case deviation of the answer occurs when the errors are either $+a$ and $-b$ or $-a$ and $+b$. In either case, the maximum error will be $(a + b)$.

In the case of the quotient, A/B, the worst-case deviation of the answer occurs when the errors have opposite sign, either $+a$ and $-b$ or $-a$ and $+b$. In either case, the maximum size of the relative error will be $(a/A + b/B)$.

The results for the operations of addition and multiplication are the same as before. In summary, **maximum indeterminate errors propagate** according to the following rules:

> **Addition and subtraction rule for indeterminate errors.**
> **The absolute indeterminate errors add.**
> **Product and quotient rule for indeterminate errors.**
> **The relative indeterminate errors add.**

A consequence of the product rule is this:

Power rule for indeterminate errors. When a quantity Q is raised to a power, P, the relative error in the result is P times the relative error in Q. This also holds for negative powers, i. e. the relative error in the square root of Q is one half the relative error in Q.

These rules apply **only** when combining **independent** errors, that is, individual errors which are not dependent on each other in size or sign.

It can be shown that these rules also apply sufficiently well to errors expressed as average deviations. The one drawback to this is that the error estimates made this way are still over conservative in that they do not fully account for the tendency of error terms associated with independent errors to offset each other. This, however, would be a minor correction of little importance in our work in this course.

Error propagation rules may be derived for other mathematical operations as needed. For example, the rules for errors in trig functions may be derived by use of trig identities, using the approximations: $\sin \beta = \beta$ and $\cos \beta = 1$, valid when β is small. Rules for exponentials may be derived also.

When mathematical operations are combined, the rules may be successively applied to each operation, and an equation may be algebraically derived which expresses the error in the result in terms of errors in the data. Such an equation can always be cast into **standard form** in which each error source appears in only one term. Let Δx represent the error in x, Δy the error in y, etc. Then the error Δr in any result R, calculated by any combination of mathematical operations from data values X, Y, Z, etc. is given by:

$$\Delta r = c_x \Delta x + c_y \Delta y + c_z \Delta z \cdots$$

This may always be algebraically rearranged to:

$$\frac{\Delta r}{R} = C_x \frac{\Delta x}{X} + C_y \frac{\Delta y}{Y} + C_z \frac{\Delta z}{Z} \cdots$$

The coefficients c_x and C_x etc. in each term are extremely important because they along with the sizes of the errors, determine how much each error affects the result. **The relative size of the terms of this equation shows us the relative importance of the error sources.** It's not the relative size of the errors (Δx, Δy, etc), but the relative size of the error terms which tells us their relative importance.

If this error equation was derived from the **determinate-error rules**, the relative errors in the above might have + or − signs. The coefficients may also have + or − signs, so the terms themselves may have + or − signs. It is therefore possible for terms to offset each other.

If this error equation was derived from the **indeterminate error rules**, the error measures appearing in it are inherently positive. The coefficients will turn out to be positive also, so terms cannot offset each other.

It is convenient to know that the indeterminate error equation may be obtained directly from the determinate-error equation by simply choosing the worst-case, i. e. , by taking the absolute value of every term. This forces all terms to be positive. This step is only done **after** the determinate-error equation has been fully derived in standard form.

The error equation in standard form is one of the most useful tools for experimental design and analysis. It should be derived (in algebraic form) even before the experiment is begun, as a guide to experimental strategy. It can show which error sources dominate, and which are negligible, thereby saving time one might spend fussing with unimportant considerations. It can suggest how the effects of error sources might be minimized by appropriate choice of the sizes of variables. It can tell you how good a measuring instrument you need to achieve a desired accuracy in the results.

The student who neglects to derive and use this equation may spend an entire lab period using instruments, strategy, or values insufficient to the requirements of the experiment. And he may end up without the slightest idea **why** the results were not as good as they ought to have been.

A final comment for those who wish to use standard deviations as indeterminate error measures: Since the standard deviation is obtained from the average of squared deviations, then each term of the equation (both sides) must be squared:

$$(\Delta r/R)^2 = (C_x)^2 (\Delta x/X)^2 + (C_y)^2 (\Delta y/Y)^2 + (C_z)^2 (\Delta z/Z)^2$$

This rule is given here without proof.

Example 1: A student finds the constant acceleration of a slowly moving object with a stopwatch. The equation used is $s = (1/2) at^2$. The time is measured with a stopwatch, the distance, s, with a meter stick.

$$s = 2 \pm 0.005 \text{ meter. This is } 0.25\%.$$
$$t = 4.2 \pm 0.2 \text{ second. This is } 4.8\%.$$

What is the acceleration and its estimated error?

We'll use capital letters for measured quantities, lower case for their errors. Solve the equation for the result, $A = 2S/T^2$. Its indeterminate-error equation is:

$$\frac{\Delta a}{A} = 2\frac{\Delta t}{T} + \frac{\Delta s}{S}$$

The factor of 2 in the time term causes that term to dominate, for application of the rule for errors in quantities raised to a power causes the 4.8% error in the time to be doubled, giving over 9.5% error in T^2. The 1/4 percent error due to the distance measurement is clearly negligible compared to the 9.5% error due to the time measurement, so the result (the acceleration) is written: $A = 0.23 \pm 0.02$ m/s^2.

Example 2: A result is calculated from the equation $R = (G + H)/Z$, the data values being:

$$G = 20 \pm 0.5$$
$$H = 16 \pm 0.5$$
$$Z = 106 \pm 1.0$$

The ± symbol tells us that these errors are indeterminate. The calculation of R requires both addition and division, and gives the value $R = 3.40$. The error calculation requires both the addition and multiplication rule, applied in succession, in the same order as the operations performed in calculating R itself.

The addition rule says that the absolute errors in G and H add, so the error in the numerator is $1.0/36 = 0.28$.

The division rule requires that we use **relative** (fractional errors). The relative error in the numerator is $1.0/36 = 0.028$. The relative error in the denominator is $1.0/106 = 0.0094$. The relative error in the denominator is added to that of the numerator to give 0.0374, which is the relative error in R.

If the **absolute** error in R is required, it is $(0.0374) R = 0.0136$. The result, with its error, may be expressed as: $R = 0.338 \pm 0.014$.

Example 3: Write a determinate-error equation for example 2, $R = (G + H)/Z$.

We follow the same steps, but represent the errors symbolically. Let N represent the numerator, $N = G + H$. The determinate error in N is then $\Delta g + \Delta h$. The relative error in the numerator is $(\Delta g + \Delta h)/N$. The relative error in the denominator is $\Delta z/Z$. The relative error in R is then:

$$\frac{\Delta r}{R} = \frac{G}{G+H} - \frac{\Delta g}{G} + \frac{H}{G+H} - \frac{\Delta h}{H} - \frac{\Delta z}{Z}$$

This equation is in standard form; each error, g, h, and z appears in only one term, that term representing that error's contribution to the error in R.

Example 4: Derive the indeterminate error equation for this same formula, $R = (G + H)/Z$.

Here's where our previous work pays off. Look at the determinate error equation of example 3 and rewrite it for the **worst case** of signs of the terms. That's equivalent to making all of the terms of the standard form equation positive:

$$\frac{\Delta r}{R} = \frac{G}{G+H} - \frac{\Delta g}{G} + \frac{H}{G+H} - \frac{\Delta h}{H} + \frac{\Delta z}{Z}$$

Example 5: Rework example 2, this time using the indeterminate error equation obtained in example 4.

Putting in the values:

$$\Delta r/R = 20/(20+16) - 0.5/20 + 16/(20+16) - 0.5/16 + 1/106$$
$$= 0.014 + 0.014 + 0.0094 = 0.0374$$

This is less than 4%.

Example 6: A result, R, is calculated from the equation $R = (G+H)/Z$, with the same data values as the previous example. After the experiment is finished, it is discovered that the value of Z was 0.05 too small because of a systematic error in the measuring instrument. The result was obtained from averaging large amounts of data, and the task of recalculating a correction to each value is daunting. But that's not necessary use this information to correct the result.

Look at the determinate error equation:

$$\frac{\Delta r}{R} = \frac{G}{G+H} - \frac{\Delta g}{G} + \frac{H}{G+H} - \frac{\Delta h}{H} - \frac{\Delta z}{Z}$$

The error -0.05 in Z represents a relative error of $-0.05/106$ in Z. Assuming zero determinate error in G and H, we have:

$$\Delta r/R = -(\Delta z/Z) = -(-0.05/106)$$

So: $\quad \Delta r = (0.05/106)(0.338) = 0.0001594$

Example 7: The density of a long copper rod is to be obtained. Its length is measured with a meter stick, its diameter with micrometer calipers, and its mass with an electronic balance.

$L = 60.0 \pm 0.1$ cm $\quad (0.17\%)$
$D = 0.632 \pm 0.002$ cm $\quad (0.32\%) \quad$ [The error in D^2 is therefore 0.64%]
$m = 16.2 \pm 0.1$ g $\quad (0.0060\%)$

The cross sectional area is $pr^2 = pD^2/4$. So the density is $m/V = 4m/LpD^2$. The relative error in the result (the density) should be no more than $(0.17\% + 0.64\% + 0.0060\% = 0.82\%)$ or about 0.82%. This is written as: density = (8.61 ± 0.07) g/cm^3.

A reference book gives 8.87 g/cm^3 as the density of copper. The experimental discrepancy is 0.26, indicating that something is wrong. The student who took this data may have blundered in a measurement. Maybe the material wasn't pure copper, but a copper alloy. If it is a measurement blunder, the diameter measurement is the most likely suspect.

All indeterminate/random errors may be eliminated, **when they are recognized**! But that's what makes them so troublesome. They may not be suspected until the final results are calculated and found to disagree with theory or with "book values". This is why it is important to do at least a rough calculation of all experimental results **before leaving lab.** Determinate errors and outright blunders may then be detected before it is too late to repeat the experiment.

2.2.2 Determinate Errors/Systematic Errors

The terms determinate error and systematic error are synonyms. "Systematic" means that when the measurement of a quantity is repeated several times, the error has the same size and algebraic sign for every measurement. "Determinate" means that the size and sign of the errors are determinable.

A common cause of determinate error is miscalibrated scale, instrumental or procedural bias. For example, a miscalibrated scale or instrument, a color-blind observer matching colors. This can happen even with such a simple instrument as a meter stick. The millimeter divisions may vary in size. The end of the stick may have been sawed off inaccurately so that a fraction of the first millimeter is "lost". This error is easily eliminated by avoiding making measurements from the end of the stick; start at 10 cm instead, and subtract 10 cm from the reading.

Another cause is an outright experimental blunder. **Examples**: using an incorrect value of a constant in the equations, using the wrong units, reading a scale incorrectly.

Every effort should be made to minimize the possibility of these errors, by careful calibration of the apparatus and by use of the best possible measurement techniques.

Determinate errors can be more serious than indeterminate errors for three reasons.

(1) There is no sure method for discovering and identifying them just by looking at the experimental data.
(2) Their effects can NOT be reduced by averaging repeated measurements.
(3) A determinate error has the same size and sign for each measurement in a set of repeated measurements, so there is no opportunity for positive and negative errors to offset each other.

Propagation of Determinate Errors

The importance of estimating data errors is due to the fact that data errors propagate through the calculations to produce errors in results. **It is the size of a data error's effect on the results which is most important.** Every effort should be made to determine reasonable error estimates for every important experimental result.

We illustrate how errors propagate by first discussing how to find the amount of error in results by considering how data errors propagate through simple mathematical operations. We first consider the case of **determinate errors**: those that have known sign.

In this way we will discover certain useful rules for error propagation, we'll then be able to modify the rules to apply to other error measures and also to indeterminate errors.

We are here developing the mathematical rules for "finite differences", the algebra of numbers which have relatively small variations imposed upon them. The finite differences are those variations from "true values" caused by experimental errors.

Suppose that an experimental result is calculated from the sum of two data quantities A and B. For this discussion we'll use Δa and Δb to represent the errors in A and B respectively. The data quantities are written to explicitly show the errors:

$$(A + \Delta a) \text{ and } (B + \Delta b)$$

We allow that Δa and Δb may be either positive or negative, the signs being "in" the symbols "Δa" and "Δb." But we must emphasize that we are here considering the case where the signs of a and b are determinable, and we know what those signs are positive, or negative.

The result of adding A and B to get R is expressed by the equation: $R = A + B$. With the errors explicitly included, this is written:

$$(A + \Delta a) + (B + \Delta b) = (A + B) + (\Delta a + \Delta b)$$

The result with its error, Δr, explicitly shown, is $(R + \Delta r)$:

$$(R + \Delta r) = (A + B) + (\Delta a + \Delta b)$$

The error in R is therefore: $\Delta r = \Delta a + \Delta b$.

We conclude that the determinate error in the sum of two quantities is just the sum of the errors in those quantities. You can easily work out for yourself the case where the result is calculated from the **difference** of two quantities. In that case the determinate error in the result will be the difference in the errors. Summarizing:

- **Sum rule for determinate errors. When two quantities are added, their determinate errors add.**
- **Difference rule for determinate errors. When two quantities are subtracted, their determinate errors subtract.**

Now let's consider a result obtained by multiplication, $R = AB$. With errors explicitly included:

$$(R + \Delta r) = (A + \Delta a)(B + \Delta b) = AB + \Delta a B + A \Delta b + \Delta a \Delta b$$

or: $\Delta r = \Delta a B + A \Delta b + \Delta a \Delta b$

This doesn't look promising for recasting as a simple rule. However, when we express the errors in relative form, things look better. If the error Δa is small relative to A, and Δb is small relative to B, then $(\Delta a \Delta b)$ is certainly small relative to AB, as well as small compared to $(\Delta a B)$ and $(A \Delta b)$. Therefore we neglect the term $(\Delta a \Delta b)$, since we are interested only in error estimates to one or two significant figures. Now we express the relative error in R as

$$\frac{\Delta r}{R} = \frac{\Delta a B + \Delta b A}{AB} = \frac{\Delta a}{A} + \frac{\Delta b}{B}$$

This gives us a very simple rule:

- **Product rule for determinate errors. When two quantities are multiplied, their relative determinate errors add.**

A similar procedure may be carried out for the quotient of two quantities, $R = A/B$.

$$\frac{\Delta r}{R} = \frac{\left(\frac{A + \Delta a}{B + \Delta b} - \frac{A}{B}\right)}{\frac{A}{B}} = \frac{\frac{(A + \Delta a)B - A(B + \Delta b)}{(B + \Delta b)B}}{\frac{A}{B}}$$

$$= \frac{(A + \Delta a)B - A(B + \Delta b)}{A(B + \Delta b)} = \frac{AB + \Delta a \cdot B - AB - A \cdot \Delta b}{(B + \Delta b)A}$$

$$= \frac{\Delta a \cdot B - A \cdot \Delta b}{AB} = \frac{\Delta a}{A} - \frac{\Delta b}{B}$$

The approximation made in the next to last step was to neglect Δb in the denominator, which is valid if the relative errors are small. So the result is:

Quotient rule for determinate errors. When two quantities are divided, the relative determinate error of the quotient is the relative determinate error of the numerator minus the relative determinate error of the denominator.

A consequence of the product rule is:

Power rule for determinate errors. When a quantity Q is raised to a power, P, the relative determinate error in the result is P times the relative determinate error in Q. This also holds for negative powers, i. e. the relative determinate error in the square root of Q is one half the relative determinate error in Q.

One illustrative practical use of determinate errors is the case of correcting a result when you discover, after completing lengthy measurements and calculations, that there was a determinate error in one or more of the measurements. Perhaps a scale or meter had been miscalibrated. You discover this, and fine the size and sign of the error in that measuring tool. Rather than repeat all the measurements, you may construct the determinate-error equation and use your knowledge of the miscalibration error to correct the result. As you will see in the following sections, you will usually have to construct the error equation anyway, so why not use it to correct for the discovered error, rather than repeating all the calculations?

The Nature of Determinate Errors

Most of the preceding discussion of errors was devoted to indeterminate errors. This should not be taken to imply that determinate errors are not important. They are a constant source of trouble in experiments, and their detection and elimination may occupy a major portion of the experimenter's time.

While indeterminate errors show up clearly as scatter in data, determinate errors cannot be detected merely by a mathematical analysis of the data. A determinate error, if present, has constant magnitude and sign for all measurements of a particular quantity. Taking many measurements does not help either to detect or to eliminate the error.

Causes of determinate error were listed:

(1) **Miscalibration of apparatus.** This can be removed by checking the apparatus against a standard.
(2) **Faulty observation.** This is avoidable, and therefore should not be cited as a source of error in any well-performed experiment.
(3) **Unnoticed outside influences.** These are also avoidable, but may be difficult to discover.

In principle all determinate errors are avoidable, but their presence is not always obvious. The first hint of a determinate error may come when experimental results are found to be inconsistent with each other by amounts larger than predicted by the indeterminate-error analysis. Even when only one result is obtained, it may be inconsistent with results obtained by other experimenters or with previously established theory, indicating a possible determinate error.

In the elementary lab the problem usually shows up as a discrepancy between the experimental value and the "textbook" value. If the discrepancy is much larger than the indeterminate-error analysis predicts, it cannot be attributed to those error sources included in that analysis. One may suspect a blunder, and should then do whatever is necessary to identify it and conclusively show that it was the source of the trouble.

The cause may be an unrecognized determinate error. This should not be the end of the story, but rather the beginning of a thorough experimental search for the cause of the determinate error, and a demonstration that elimination of the suspected cause improves the result. Until this is done, any speculation about the cause of a "bad" result is only guesswork.

The physical or psychological causes of determinate error are, in principle, measurable. But if the cause was not suspected, the experimenter probably did not take the necessary measurements. One does not usually measure **everything**! The usual procedure is to design the apparatus so that unwanted influences are negligible. If this does not eliminate determinate errors, one then searches for them, or redesigns the experiment.

Psychological causes of determinate error, such as observer bias, may be extremely subtle, especially in observations of color, brightness, shape, behavior of fast-moving objects, etc. In physics we generally try to avoid such observations by using measuring instruments which require only **reproducible** visual measurements, such as reading a stationary needle on a meter scale, or geometric measurements on a photograph. Additional insurance of accuracy is provided by common sense, self-awareness, and gaining practice and experience in observational technique.

One should not regard any deviation from theory as being due to a determinate error. There is the possibility that the theory may be imperfect. This is one way deficiencies in a theory are discovered——when the experiment doesn't agree with theory.

Detection of Determinate Errors

No fixed rules can be given for tracking down determinate errors. Rather we will give one example of a general approach which is useful in some situations.

Assume that an experiment is performed which required taking data on quantities A, B, C, and D, used in the calculation of a result, R. The experiment is repeated a number of times, and many values of R are calculated, but the experimental scatter of the values of R is found to be larger than the predicted indeterminate error.

After thorough analysis of the experimental procedure, someone suspects that quantity E, not previously recorded, might be influencing the experiment. So the experiment is performed again, this time taking data on E (and perhaps on some other suspected quantities, F, G, etc.).

Now it is noticed that the largest values of R occur when E is smallest, and vice versa. To check the relation of R to E, a graph is constructed of R vs. E, see Figure 1.

The error bars on the data points are those predicted by the indeterminate error analysis. The graph clearly shows that R is not constant, for the average R (dashed line) is not as good a fit as the slanting line (dotted line). It also shows the trend of R against E (the slanting dotted line), but more data might be required to determine the precise nature of this relation.

Finally one might try to hold E constant, or try to eliminate its influence entirely, to see if constant values of R result. One should at least show, by applying known theory, that the assumed mechanism causing the determinate error can actually give rise to a discrepancy of the size and behavior observed.

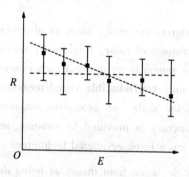

Figure 1 The relation of R to E

2.3 Measurement Uncertainty

Each instrument has an inherent amount of uncertainty in its measurement. Even the most precise measuring device cannot give the actual value because to do so would require an infinitely precise instrument. A measure of the **precision** of an instrument is given by its uncertainty. **As a good rule of thumb, the uncertainty of a measuring device is 20% of the least count.** Recall that the least count is the smallest subdivision given on the measuring device. The uncertainty of the measurement should be given with the actual measurement, for example, 41.64 cm ± 0.02 cm.

Here are some **typical** uncertainties of various laboratory instruments:

- Meter stick: ± 0.02 cm
- Venire caliper: ± 0.01 cm
- Triple-beam balance: ± 0.02 g
- Graduated cylinder: 20% of the least count

Here's an example. The uncertainty of all measurements made with a meter stick whose smallest division (or least count) is one millimeter is 20% of 1 mm or 0.02 cm. Say you use that meter stick to measure a metal rod and find that the rod is between 10.2 cm and 10.3 cm. You may think that the rod is closer to 10.2 cm than it is to 10.3 cm, so you make your **best guess** that the rod is 10.23 cm in length. Since the uncertainty in the measurement is 0.02 cm, you would report the length of the metal rod to be 10.23 ± 0.02 cm (0.1023 ± 0.0002 m).

When a quantity is graphed, it is common for the uncertainty of that quantity to be represented by error bars.

No measurement is perfectly accurate or exact. Many instrumental, physical and human limitations cause measurements to deviate from the "true" values of the quantities being measured. These deviations are called "experimental uncertainties", but more commonly the shorter word "error" is used.

What is the "true value" of a measured quantity? We can think of it as the value we'd measure if we somehow eliminated all error from instruments and procedure. This is a natural enough concept, and a useful one, even though at this point in the discussion it may sound like circular logic.

We can improve the measurement process, of course, but since we can never eliminate measurement errors entirely, **we can never hope to measure true values.** We have only introduced the concept of **true value** for purposes of discussion. When we specify the "error" in a quantity or result, we are giving an estimate of how much that measurement is likely to deviate from the true value of the quantity. This estimate is far more than a guess; for it is founded on a physical analysis of the measurement process and a mathematical analysis of the equations which apply to the instruments and to the physical process being studied.

A measurement or experimental result is of little use if nothing is known about the probable size of its error. We know nothing about the reliability of a result unless we can estimate the probable sizes of the errors and uncertainties in the data which were used to obtain that result.

That is why it is important for students to learn how to determine quantitative estimates of the nature and size of experimental errors and to predict how these errors affect the reliability of the final result. The following discussion is designed to make the student aware of some common types of errors and some simple ways to quantify them and analyze how they affect results.

The experimental error can be expressed in several standard ways:

1. Limits of Error

Error limits may be expressed in the form $Q \pm \Delta Q$ where Q is the measured quantity and ΔQ is the magnitude of its limit of error. This expresses the experimenter's judgment that the "true" value of Q lies between $Q - \Delta Q$ and $Q + \Delta Q$ this entire interval within which the measurement lies is called the **range of error**. Manufacturer's performance guarantees for laboratory instruments are often expressed this way.

2. Average Deviation

This measure of error is calculated in this manner: First calculate the mean (average) of a set of successive measurements of a quantity, Q. Then find the magnitude of the deviations of each measurement from the mean. Average these magnitudes of deviations to obtain a number called the **average deviation** of the data set. It is a measure of the dispersion (spread) of the measurements with respect to the mean value of Q, that is, of how far a typical measurement is likely to deviate from the mean. But this is not quite what is needed to express the quality of the mean itself. We want an estimate of how far the mean value of Q is likely to deviate from the "true" value of Q. The appropriate statistical estimate of this is called the **average deviation of the mean.** To find this rigorously would involve us in the theory of probability and statistics. We will state the result without proof.

For a set of n measurements Q_i whose mean value is $\langle Q \rangle$, the average deviation of the mean (A. D. M.) is:

Average deviation of the mean $\dfrac{\sum_{n-1}^{n} |Q_i - \langle Q_i \rangle|}{(n-1)\sqrt{n}}$

The vertical bars enclosing an expression mean "take the absolute value" of that expression. That means that if the expression is negative, make it positive.

If the A. D. M. is quoted as the error measure of a mean, $\langle Q \rangle_{exp}$, this is equivalent to saying that the probability of $\langle Q \rangle_{exp}$ lying within one A. D. M. of the "true" value of Q, Q_{true}, is 58%, and the odds against it lying outside of one A. D. M. are 1.4 to 1.

As a rough rule of thumb, the probability of being within three A. D. M. (on either side) of the true value is nearly 100% (actually 98%). This is a useful relation for converting (or comparing) A. D. M. to limits of error.

3. Standard Deviation of the Mean

With any experiment it is important to properly display the precision with which each measurement is made. No measurement is absolutely precise. For example, it is impossible to measure the exact length of an object. We might measure the length as 1.23 cm, but this does not mean that the actual measurement is 1.23000000... cm! We must carefully describe how precise our measurement is. An experimental value of 1.23 ± 0.10 cm is less precise than a measurement of 1.23 ± 0.01 cm. The ± term gives the measure of the precision of the measurement. The accuracy of the value is given either by percent error or percent difference.

To find the uncertainty in our measurements, we will often calculate the **standard deviation**, or σ, of the measured value. Standard deviation is a measure of the variation of N data points ($x_1 \ldots x_N$) about an average value, $\overline{X}$ and is typically called the uncertainty in a measured result.

The **average or mean value**, $\overline{X}$ of a set of N measurements is

$$\overline{X} = \frac{x_1 + x_2 + x_3 + \cdots + x_N}{N} = \frac{1}{N}\sum_{i=1}^{N} x_i \tag{1}$$

Once the mean value of the measurements is determined, it is helpful to define how much the individual measurements are scattered around about the mean. The **deviation**, d_i, of any measurement, x_i, from the mean is given by $d_i = x_i - \overline{X}$.

Since the deviation may be either positive or negative, it is often more useful to use the **mean deviation**, or $\overline{d}$, to determine the uncertainty of the measurement. This is found by averaging the absolute deviations, $|d_i| = |x_i - \overline{x}|$; that is,

$$\overline{d} = \frac{|d_1| + |d_2| + |d_3| + \cdots + |d_N|}{N} = \frac{1}{N}\sum_{i=1}^{N} |d_i| \qquad (2)$$

It is common to report the experimental value, E_x, of a measurement as $E_x = \overline{x} \pm \overline{d}$ where $\overline{d}$, gives the measure of the precision of the measurement. To avoid the use of absolute values we can use the square of the deviation, d_i^2, to more accurately determine the uncertainty of our measurement. The standard deviation, σ, (sometimes called the **root-mean square, rms**) is given by

$$\sigma = \sqrt{\frac{(x_1 - \overline{x})^2 + (x_2 - \overline{x})^2 + \cdots + (x_N - \overline{x})^2}{N}}$$

$$= \sqrt{\frac{d_1^2 + d_2^2 + \cdots + d_N^2}{N}}$$

$$= \sqrt{\frac{1}{N}\sum_{i=1}^{N}(x_i - \overline{x})^2}$$

$$= \sqrt{\frac{1}{N}\sum_{i=1}^{N} d_i^2} \qquad (3)$$

It can be shown that **for a small number of measurements**, where N in Eq. (3) is replaced by $N - 1$.

Finally, the experimental result, E_x, can then be written as $E_x = \overline{x} \pm \sigma$, where σ, gives the measure of the precision of the measurement. Often scientists use the value of the standard deviation to serve as their data's Error bars. Notice the standard deviation is always positive and has the same units as the mean value. It can be shown that there is a 68% likelihood that an individual measurement will fall within one standard deviation ($\pm \sigma$) of the true value. Furthermore, it can be shown that there also exists a 95% likelihood that an individual measurement will fall within two standard deviations ($\pm 2\sigma$) of the true value, and a 99.7% likelihood that it will fall within ($\pm 3\sigma$) of the true value (Figure 1).

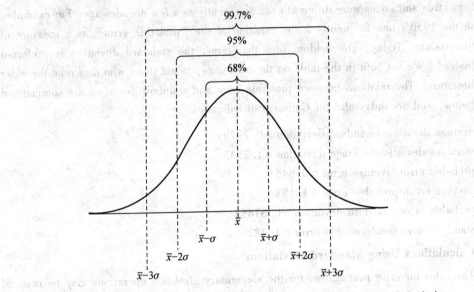

Figure 1 The relation ship between individual measurement and standard deviations
From http://www.everything maths.co.za/grade

Expand the summand in the standard deviation:

$$\sum (x_i - \langle x \rangle)^2 = \sum (x_i^2 - 2x_i\langle x \rangle + \langle x \rangle^2)$$
$$= \sum x_i^2 - 2\langle x \rangle \sum x_i + n\langle x \rangle^2$$
$$= \sum x_i^2 - 2\langle x \rangle n\langle x \rangle + n\langle x \rangle^2$$
$$= \sum x_i^2 - n\langle x \rangle^2$$

So, **for a small number of measurements**, the standard deviation becomes

$$\sigma = \sqrt{\frac{\sum x_i^2 - n\langle x \rangle^2}{n-1}} \qquad (4)$$

Many electronic calculators have a built-in routine which allows you to enter the x_i values in succession. When this is done, the calculator has accumulated the sum of those values in one memory register, the sum of the squares of the values in another register, and may even have counted the entries and stored n in a register. These stored values are then easily recalled to calculate the standard deviation.

A similar procedure can be used for the rms deviation.

Standards and styles were different even as recently as a few decades ago. For example, in the 1950's one frequently found mention of the "probable error" as a measure of uncertainty. Today, one seldom sees that term, the standard deviation is preferred instead. We list both in the table on the next page, to aid those who may read the older literature. The relations between probable error and standard deviation are summarized below, and are only valid for Gaussian distributions.

average deviation/standard deviation = 0.7979;
standard deviation/average deviation = 1.2533;
probable error/average error = 0.8453;
average error/probable error = 1.183
probable error/standard deviation = 0.6745;
standard deviation/probable error = 1.4826

Calculations Using Standard Deviations

The rules for error propagation for the elementary algebraic operations may be restated to apply when standard deviations are used as the error measure for random (indeterminate) errors:

- **When independently measured quantities are added or subtracted, the standard deviation of the result is the square root of the sum of the squares of the** *standard deviations* **of the quantities.**
- **When independently measured quantities are multiplied or divided, the relative (fractional or percent) standard deviation of the result is the square root of the sum of the squares of the** *relative standard deviations* **of the quantities.**

Are Standard Deviations Better? Too many elementary laboratory manuals stress the standard deviation as the one standard way to express error measures. However, one can find, from standard statistical theory that when very few measurements are made, the error estimates themselves will have low precision. The uncertainty of an error estimate made from n pieces of data is $100/\sqrt{2(n-1)}$ percent.

So we'd have to average 51 independent values to obtain a 10% error in the determination of the error. We would need 5000 measurements to get an error estimate good to 1%. If only 10 measurements were made, the uncertainty in the standard deviation is about 24%. This is why we have continually stressed that error estimates of 1 or 2 significant figures are sufficient when data samples are small.

This is just one reason why the use of the standard deviation in elementary laboratory is seldom justified. How often does one take more than a few measurements of each quantity? Does one even take enough measurements to determine the nature of the error distribution? Is it Gaussian, or something else? One usually doesn't know. If it isn't close to Gaussian, the whole apparatus of the usual statistical error rules for standard deviation must be modified. But the rules for maximum error, limits of error, and average error are sufficiently conservative and robust that they can still be reliably used even for small samples.

Standard Methods for Expressing Error

1. Absolute Error

Uncertainties may be expressed as **absolute** measures, giving the size of the a quantity's uncertainty in the same units in the quantity itself.

Example 1: A piece of metal is weighed a number of times, and the average value obtained is: $M = 34.60$ g. By analysis of the scatter of the measurements, the uncertainty is determined to be $m = 0.07$ g. This **absolute** uncertainty may be included with the measurement in this manner: $M = (34.60 \pm 0.07)$ g.

The value 0.07 after the ± sign in this example is the estimated absolute error in the value 34.60.

2. Relative (or Fractional) Error

Uncertainties may be expressed as **relative** measures, giving the ratio of the quantity's uncertainty to the quantity itself. In general:

Relative error = absolute error in a measurement/size of the measurement

$$\text{Percent Error} = \frac{|\text{measured} - \text{actual}|}{\text{actual}} \times 100\%$$

Example 2: In the previous example, the uncertainty in $M = 34.6$ g was $m = 0.07$ g. The relative uncertainty is therefore: $m/M = 0.07 \text{ g}/34.6 \text{ g} = 0.002 = 0.20\%$.

Sometimes, we will compare the results of two measurements of the same quantity. For instance, we may use two different methods to determine the speed of a rolling body. In this case, since there is **not** one accepted value for the speed of a rolling body, we will use the **percent difference** to determine the similarity of the measurements. This is found by dividing the absolute difference of the two measured values by their average, or

$$\text{Percent Difference} = \frac{|\text{measured}_1 - \text{measured}_2|}{\left(\frac{\text{measured}_1 + \text{measured}_2}{2}\right)} \times 100\%$$

[**Exercises**]

1. What percent of the measurements fall within the width at half height of a Gaussian curve?

2. A set of measurements of a quantity is

878	849	804	755
816	833	781	735
964	795	817	807
862	801	778	810
778	799	819	797

   Find the means, average deviations, and standard deviations for 1) each of the four groups, and 2) the whole group of twenty.

3. Graph the distribution of Exercise 2.

[**Appendix**]

Error Propagation

The Error Analysis should include measurement uncertainties for all instruments used during the experiment. It should list any assumptions made during the experiment, along with arguments for or against their validity in the context of the experiment. Any sources of error found during the experiment should be discussed, including a qualitative statement regarding its possible impact on the results of the experiment. Finally, any specific methods used during the experiment to control or eliminate error should be listed, again including a qualitative statement regarding the possible impact on the results of the experiment.

To totally omit a concern for errors does a disservice to the student and leaves the false notion that "getting the right answer" represents the major objective of laboratory work. If students measure the "goodness" of the experiment by how close they come to the "textbook value", then we have perpetuated an attitude that runs counter to good scientific method.

To talk about sophisticated statistical measures of error has little value when one hasn't taken enough data to even know what kind of error distribution the measurements have.

One might as well use the crude "maximum error" measure, even with its limitations.

We can, and should, emphasize propagation of errors, and the error propagation equation. No matter what the error distribution, or what fancy, or crude, error measures one uses, error propagation equations describe the effect that errors have on results. The error equations have a far greater importance than merely calculating the error in a result. When a student uses the error equation to optimize the experimental procedure and thereby minimize errors in results, this improves the experiment and its results no matter what error measure one happens to use. This approach has value even if the student has never heard of "standard deviation".

The error propagation equation serves to guide the experimental strategy, identifying those variables that most affect the error, and it shows what must be done to attain a desired precision in the result. The error equation may sometimes guide the experimenter in the choice of the sizes of variables to produce the best results.

Once the student learns the technique of error propagation analysis, and its use becomes habitual, a solid foundation has been built for later work, and for conceptually understanding the more sophisticated mathematical analyses of error.

Attention to error propagation encourages the student to think, in a critical manner, about the entire experiment, from overall strategy to the minutest detail. This may represent the most valuable benefit from concern for error analysis in the elementary laboratory.

When dealing with uncertainties based on a large collection of numbers the manipulation of measured quantities and the error associated with each quantity will contribute to the error in the final answer. The following formulae are a good approximation of the error and become increasingly accurate as the number of measurements increase or when the cross terms between the contributing errors are reasonably small.

Process	Value	Uncertainty
Average	$\bar{x} = \dfrac{x_1 + x_2 + x_3 + \cdots + x_N}{N}$	$\sigma_x = \sqrt{\dfrac{(x_1 - \bar{x})^2 + (x_2 - \bar{x})^2 + \cdots + (x_N - \bar{x})^2}{N - 1}}$
Addition	$\bar{z} = \bar{x} + \bar{y}$	$\sigma_z = \sqrt{(\sigma_x)^2 + (\sigma_y)^2}$
Subtraction	$\bar{z} = \bar{x} - \bar{y}$	$\sigma_z = \sqrt{(\sigma_x)^2 + (\sigma_y)^2}$
Multiplication	$\bar{z} = \bar{x} \cdot \bar{y}$	$\sigma_z = \bar{z} \cdot \sqrt{\left(\dfrac{\sigma_x}{\bar{x}}\right)^2 + \left(\dfrac{\sigma_y}{\bar{y}}\right)^2}$
Division	$\bar{z} = \bar{x}/\bar{y}$	$\sigma_z = \bar{z} \Big/ \sqrt{\left(\dfrac{\sigma_x}{\bar{x}}\right)^2 + \left(\dfrac{\sigma_y}{\bar{y}}\right)^2}$

Error Propagation Rules
for Elementary Operations and Functions

Let R be the result of a calculation, without consideration of errors, and ΔR be the error (uncertainty) in that result. Determinate errors have determinable sign and constant size. Indeterminate errors have unpredictable size and sign, with equal likelihood of being + or −.

Rules for Elementary Operations (Determinate Errors)

Sum rule: When $R = A + B$ then $\Delta R = \Delta A + \Delta B$
Difference rule: When $R = A - B$ then $\Delta R = \Delta A - \Delta B$
Product rule: When $R = AB$ then $(\Delta R)/R = (\Delta A)/A + (\Delta B)/B$
Quotient rule: When $R = A/B$ then $(\Delta R)/R = (\Delta A)/A - (\Delta B)/B$
Power rule: When $R = A^n$ then $(\Delta R)/R = n(\Delta A)/A$
  or $(\Delta R) = nA^{n-1}(\Delta A)$

Memory clues: When quantities are added (or subtracted) their absolute errors add (or subtract). But when quantities are multiplied (or divided), their relative fractional errors add (or subtract).

We can also collect and tabulate the results for commonly used elementary functions. Note: Where $\Delta\tau$ appears, it must be expressed in **radians**.

Rules for Elementary Functions (Determinate Errors)

Table 1 Rules for Elementary Functions (Determinate errors)

EQUATION	ERROR EQUATION
$R = \sin\theta$	$\delta R = (\delta\theta)\cos\theta$
$R = \cos\theta$	$\delta R = -(\delta\theta)\sin\theta$
$R = \tan\theta$	$\delta R = (\delta\theta)\sec^2\theta$
$R = e^x$	$\delta R = (\delta x)e^x$
$R = e^{-x}$	$\delta R = -(\delta x)e^{-x}$
$R = \ln(x)$	$\delta R = (\delta x)/x$

Any measures of error may be converted to relative (fractional) form by using the definition of relative error. The fractional error in x is: $f_x = \dfrac{(\Delta R)x}{x}$ where $(\Delta R)x$ is the absolute error in x. Therefore $x \cdot f_x = (\Delta R) \cdot x$.

The rules for indeterminate errors are simpler.

Table 2   Rules for elementary operations (indeterminate errors)

Rules for Elementary Operations (Indeterminate Errors)		
Sum or Difference:	When $R = A + B$	then $\Delta R = \Delta A + \Delta B$
Product or Quotient:	When $R = AB$	then $(\Delta R)/R = (\Delta A)/A + (\Delta B)/B$
Power Rule:	When $R = A^n$	then $(\Delta R)/R = n(\Delta A)/A$
		or $(\Delta R) = nA^{n-1}(\Delta A)$

The indeterminate error rules for elementary functions are the same as those for determinate errors **except** that the error terms on the right are all positive.

Table 2. Rules for elementary operations (Indeterminate errors)

Rule for Elementary Operations (Indeterminate Errors)	
Sum or Difference:	If $R = A \pm B$, then $\Delta R = \sqrt{(\Delta A)^2 + (\Delta B)^2}$
Product or Quotient:	If $R = AB$ or $R = A/B$, then $(\Delta R/R)^2 = (\Delta A/A)^2 + (\Delta B/B)^2$
Power Rule:	If $R = A^n$, then $\Delta R/R = n\,\Delta A/A$

The indeterminate error rules for elementary operations are the same as those for determinate errors except that the error terms on the right are all positive.

2.4 Significant Figures

Unfortunately data errors propagate through calculations, usually producing even worse error in the results. In the following discussion we review the "rules for significant figures", a crude method for ensuring that calculated results are stated to a precision consistent with the precision of the data.

Significant figures are the number of reliably known digits used to locate a decimal point reported in a measurement. Proper use of significant figures ensures that you correctly represent the uncertainty of your measurements. E. g., scientists immediately realize that a reported measurement of 1.2345 m is much more accurate than a reported length of 1.2 m.

Now suppose that the number 3586.297 cm represents an experimental measurement, and we experimentally determined that its uncertainty was 0.2 cm. The size of the uncertainty tells us that the digits 9 and 7 are superfluous, and carry no significant information. They express an amount smaller than the uncertainty. Such digits are called insignificant. Insignificant digits can arise from mathematical calculations. Calculation devices such as electronic calculators display insignificant digits. They may also arise from reading an instrument scale beyond the inherent precision of the instrument. This terminology may be summarized with a diagram.

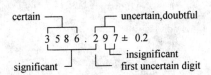

Care must be taken when determining the number of significant figures to use. Your driver's license may state that your height is 5 feet 3 inches, or 5.25 feet. Measuring a little more carefully, we may find that your height is found to be 5.257 feet. However if you said you were 5.257186 feet tall the scientific community would look upon that measurement with serious skepticism because you stated your height accurate to the nearest micron! When a measurement is properly stated in scientific notation all of the digits will be significant. For example: 0.0035 has 2 significant figures which can be easily seen when written in scientific notation as 3.5×10^{-3}. Fortunately, there are a few general guidelines that are used to determine significant figures:

Rule 1 (a): Write values of physical measurements so that the last measured

digit falls somewhere to the right of the decimal point. This may be done in either of two ways:

1) Use "scientific notation" (powers of 10).
2) Choose larger units of measurement.

Rule 1 (b): The digit representing the smallest measured scale division must be explicitly written, even if it is a zero.

Whole Numbers: Note that zeros are often place holders and are not significant. The following numbers are all represented by **three** significant digits.
- 0.00123
- 0.123
- 1.23
- 12.3
- 123

The following numbers are all represented by **one** significant digit.
- 0.005
- 0.5
- 5

The following numbers are all represented by **four** significant figures.
- 0.004001
- 0.004000
- 40.01
- 40.00
- 4321
- 432.1

Example 1: The distance between two towns is measured to the nearest 10 meters and found to be 387 220 meters. To express this correctly we may write it in one of the following ways: 387.22 kilometers (rule 1 (a), 2); 3.8722×10^5 meters (rule 1 (a), 1); 3.8722×10^7 centimeters etc. (rule 1 (a), 1).

Parts (a) and (b) of this rule tell you how to discard insignificant digits, a process called truncating. Some people get fussy about a special case that occurs, when you must truncate (discard) just one digit that happens to be a "5". Should the final result be rounded "up" or "down"? Some advocate rounding (altering) the last digit retained so that it is even. Thus "3.785" would round to "3.78" while "2.755" would round to "2.76". When many numbers are combined, this rule serves to minimize the bias introduced by consistently rounding "5" up (or rounding it down consistently). This is only important if you use many rounded numbers in calculations, which you are not likely to ever do. However, textbooks and published papers use this rule for expressing numbers, so it's good to know about it.

Rule 2: First discard all insignificant digits except the leading one. Then round off this uncertain digit. If the first uncertain digit is a 5, round it up or down as necessary to make the result even.

You could round up or down to make the result odd, so long as you adopted the same rule consistently everywhere in the calculation. This rule ensures that over many calculations you won't be introducing systematic error in one direction (up or down).

Example 2: A set of data has an average value of 3.645987 cm. The uncertainty is found to be 0.03 cm. Discard the 987. The first uncertain digit is 5. Round it down, to give the result 3.64 if you are using the "even" rule. You may object that the 5987 would suggest rounding up. But remember, we first discarded those digits 987 as insignificant, which means that digit is totally unreliable for any purpose. We must take the word "insignificant" seriously. It would be a waste of time to apply rounding rules tediously to a string of insignificant digits.

Rule 2 is a bit of "overkill", considering that significant figures rules are themselves only an approximate indication of quality of a result. Little is lost by simply discarding all insignificant digits. If you are taking a course which expects you to use this rule, you may find your results sometimes differ from the "book values" by 1. That's no "big deal".

Rule 3: Multiplication or division. Results of multiplication or division are rounded to the same number of significant digits as the least accurate data quantity.

Consider this multiplication example, in which uncertain digits are shown in **bold italics**.

$$
\begin{array}{r}
3954 \\
\times)\ \ 286 \\
\hline
23724 \\
31632 \\
7908 \\
\hline
=)\ 1139844
\end{array}
$$

Multiplying 3954 by the uncertain digit 6 gives a number in which every digit is uncertain. In the other sub multiplications, digits resulting from multiplication by the uncertain digit 4 must be counted as uncertain. Any column addition containing uncertain digits gives an uncertain result. Therefore only the first two digits of the answer are certain. The three is uncertain, and the remaining digits are completely uncertain. Therefore this result should be rounded to 114. Notice that even though the multiplicand had four significant digits, the result has only three.

Integers and Defined Quantities: Integers are assumed to have an infinite number of significant figures. For example, the 2 in $C = 2pr$, is exactly two and we can assume that the number has an infinite number of significant figures. However, the conversion factor 2.54 cm which is used to convert inches to centimeters has three significant figures.

Multiplication and Division: When multiplying or dividing numbers, the result should have only as many significant figures as the quantity with the **smallest** number of significant figures being used in the calculation. For example, with your calculator multiply 4.7 and 5.93. The calculator returns 27.871 as the answer. A common mistake students make is to record what comes out of the calculator as the correct answer. However, since 4.7 has only 2 significant figures, the result must be truncated to 2 significant figures as well. Taking all this into account and remembering to round appropriately, the result should be reported as 28.

Rule 4: Addition or subtraction. When adding or subtracting numbers, keep all figures up to the smallest quantity being used in the calculation. Find the data quantity whose last significant digit occupies the leftmost decimal place. This is the position of the last significant digit of the result.

Example 3:

```
         0.52865
        39.42
        15.1
    +)   0.02896
        ─────────
    +)  55.07761
```
, which should be rounded to 55.1.

Example 4: 3.14 + 0.00159 = 3.14*159*. It should be rounded to 3.14.

Rule 5: When other operations are encountered, analyze the computational procedure as we have illustrated above.

Example 5. Find the square root of 9.4263. The significant number is five. The square root displayed on the calculator screen is 3.070228, which gives not the slightest clue how many of the digits are significant. Obviously we need to examine this matter of uncertainties in more detail, with the goal of developing a relatively simple and reliable mathematical process for estimating the precision of results of mathematical calculations.

2.5 Methods of Data Analysis

The **graphical analysis** of data is most useful for communicating results in reports and for gaining intuitive understanding of phenomena. However, when accurate results are required, **analytic methods** are preferred over graphical methods. Here will introduce some of these methods.

The purpose of data analysis is to use all of the data to calculate one or more results. This is usually done by averaging large amounts of data. The averaging method must be carefully chosen so that it actually uses all of the data in a consistent way. The methods described below give results which are the best obtainable for the data used. Implicit in all these methods is the assumption that the individual data values have Gaussian distributions. If there is good reason to believe that the distributions are not Gaussian, modified methods are required.

Data analysis can show which quantities must be measured most precisely. It can show that some experimental designs are unsuitable for good measurements of some quantities, suggesting a search for better designs.

2.5.1 Graphing

A graph is an accurate pictorial representation of data. The accuracy of data in physics requires that graphs be made on good quality graph paper. Nearly all graphs in physics are smooth line graphs; broken line graphs and bar graphs are seldom appropriate. The style and format of a graph will depend upon its intended purpose. Three types are common in physics.

1. Pictorial Graphs

These are the kind found in mathematics and physics textbooks. Their purpose is to simply and clearly illustrate a mathematical relation. No attempt is made to show data points or errors on such a graph.

2. Display Graphs

These present the data from an experiment. They are found in laboratory reports, research journals, and sometimes in textbooks. They show the data points as well as a smooth line representing the mathematical relation.

3. Computational Graphs

These are drawn for the purpose of extracting a numerical result from the data. An example is the calculation of the slope of a straight line graph, or its intercepts.

General information about graphing:

(1) All graphs should be done by hand in pencil on graph paper.
(2) Unless instructed to do so, draw only one graph per page.
(3) The graph should use as much of the graph paper as possible. Carefully choosing the best scale is necessary to achieve this. The axes should extend beyond the first and last data points in both directions.
(4) All graphs should have a short, descriptive title or caption at the top of each graph, detailing what is being measured or clearly stating what the graph illustrates.
(5) Each axis should be clearly labeled with titles and units.
(6) Clearly label the scale of each axis. Choose scales that are convenient to plot and easy to read. Choose scales such that the graph occupies most of the page. The two scales need not have the same size units. Also, the scales need not begin at zero.
(7) Indicate the name, letter symbol and units of each variable plotted on each axis.
(8) Never connect the dots on a graph, but rather give a best-fit line or curve. A best-fit line should extend beyond the data points.
(9) The best-fit line should be drawn with a ruler or similar straight edge, and should closely approximate the trend of all the data, not any single point or group of points.
(10) The slope should be calculated from two points on the best-fit line. The two points should be spaced reasonably far apart. Data points should not be used to calculate the slope.
(11) On a linear graph, draw the rise, Δy, and run, Δx, to form a triangle with the best-fit line. Be sure to label these values and include units.
(12) The calculation of the slope, $\frac{\Delta y}{\Delta x} = \frac{y_2 - y_1}{x_2 - x_1}$, should be clearly shown on the graph itself. Units should be included, and value of the slope should be easily visible.
(13) See the sample graph below which incorporates the above requirements.

Physical Slope and Geometric Slope

Slope. When textbooks refer to the "slope" of a plotted graph line we mean the "physical slope" = $\Delta y/\Delta x$. Where Δy and Δx are expressed in the physical units of the x and y axes. This slope has physical significance in describing the physical data.

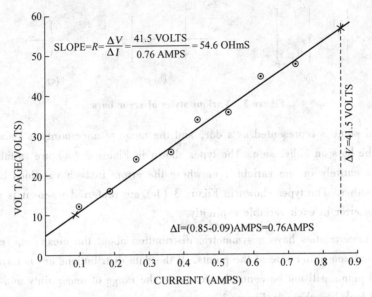

Figure 1 Experimental Resistance of A "50Ω" Resistor

Geometric slope. A line which makes a 45° angle with an axis will not necessarily have a physical slope of size 1. Some authors introduce the term "geometric slope" to describe the tilt of the line on the page. This is a ratio of lengths of the legs of the triangle, without reference to the units plotted on the axes.

There is seldom (probably never) any need to calculate the geometric slope of a line on a graph. The idea is only useful when describing the appearance of the graph on the page. One rule of graph construction states that the graph should occupy most of the page. For square graph paper this suggests a geometric slope of 45° angle. See Figure 2 for examples of good and bad choices of geometric slope.

Figure 2 Good and bad choices of geometric slope of a graph

Graphical Representation of Uncertainties

Display graphs and computational graphs should clearly show the size of the experimental uncertainties (errors) in each plotted point. There are several conventional ways to do this, the commonest being the use of error bars illustrated below (Figure 3):

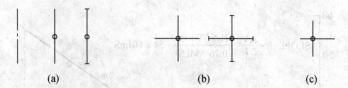

Figure 3 Various styles of error bars

The plotted point is represented as a dot, and the range of uncertainty is shown by the extent of the bars on either side. The types shown in Figure 3 (a) are suitable where the error is entirely in one variable, or where the errors in both variables have been lumped together. The types shown in Figure 3 (b) are preferred where it is necessary to show the error in each variable explicitly.

When the uncertainties have a symmetric distribution about the mean, the error bars extend equally on either side of the points. If the data distributions are not symmetric, the plotted points will not be centrally located in the range of uncertainty and the error bars might look like those in Figure 3 (c).

Error bars may not be necessary when the data points are so numerous that their scatter is clearly shows the uncertainty. In these cases error bars would clutter the graph making it difficult to interpret. Another situation where error bars are inappropriate is that the scale of the graph makes the bars very small. In this case, it may be possible to indicate the uncertainty by the size of the circle or rectangle surrounding each point.

Curve Fitting

The curve drawn through plotted data need not pass exactly through every data point. But usually the curve should pass within the uncertainty range of each point, that is, within the error bars, if the bars represent limits of error.

One principle of curve fitting is also a fundamental rule of science itself:

Assume the simplest relation consistent with the data.

We are not justified in assuming a more complex relation than can be demonstrated by the data. If a curve were drawn with detail smaller than the data uncertainty, it would be only a guess.

This rule of simplicity may also be expressed mathematically. The mathematical relations encountered in physics may often be represented by power series such as

$$y = A + Bx + Cx^2 + Dx^3 + Ex^4 + \cdots \tag{1}$$

where A, B, C, ... are constants.

For very "wiggly" curves, many terms of this equation, including high powers of x, might be required to express the equation of the relation. The simplest relations are those which contain the smallest powers of x. The simplest relations of all are

$$y = a \quad \text{or} \quad y = a + bx \tag{2}$$

which describe straight lines. Many relations in physics are, fortunately, of this form. Others only include the x^2 term, describing a parabolic curve. Note that double valued curves, sometimes encountered in physics, cannot be represented by Eq. (1).

When sizable amounts of data are taken, standard mathematical methods are available which generate the equation of the simplest curve which statistically "best fits" the data.

The student may wonder how one can be certain that the curve fitted to the data is the "correct" curve. The answer is that relations are never known with certainty. The uncertainty of available data always limits the certainty of the results. Someday someone may obtain more accurate data and be able to show that the old relations are slightly incorrect, and provide us with better ones. As data improve, so does our understanding of relations—this is the way of scientific progress. But we never should claim to know a relation better than the data allow.

Uncertainty in a Slope

One use of a computational graph is to determine the slope of a straight line. This is illustrated in Figure 4. Eight data points are shown with error bars on each. If these bars represent maximum error, any line drawn to represent this data should pass within all bars.

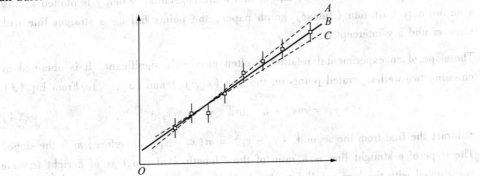

Figure 4 Fitting a curve

If the error bars represent error estimates smaller than the maximum (average deviation, standard deviation, etc.), then the fitted curve need not pass within all of the error bars, just most of them.

Even a simple "manual" curve fit with a ruler can reveal the uncertainties in the slope resulting from uncertainties in the data. Figure 4 illustrates this process.

The dotted lines A and C fall within the error bars, and represent the maximum and minimum slope one could justify from this data. The "best" value of slope might be that of solid line B.

The third point from the left seems to limit the slope the most, and would appear to be "suspect." But one ought not to "throw it out" without better reason, based on further investigation.

Graphical Analysis of Data

Graphs can be a valuable tool for determining or verifying functional relations between variables. Many special types of graph paper are available for handling the most frequently encountered relations. You are probably already familiar with linear graph paper and polar coordinate paper.

You may have purchased a packet of graph paper for this course. It includes samples of graph papers you will use in this course, and a few other types. As you read the material below, examine the corresponding papers from your packet.

Linear relations are those which satisfy the equation

$$y = mx + b \qquad (3)$$

where the variables are x and y, and m and b are constants. When y is plotted against x on ordinary Cartesian (linear) graph paper, the points fall on a straight line with slope m and a y-intercept b.

The slope of an experimental relation is often physically significant. It is obtained by choosing two well-separated points on the line (x_1, y_1) and (x_2, y_2). From Eq. (3)

$$y_1 = mx_1 + b \quad \text{and} \quad y_2 = mx_2 + b \qquad (4)$$

Subtract the first from the second. $(y_2 - y_1) = m(x_2 - x_1)$, therefore, m is the slope. The slope of a straight line is a ratio of the "lengths" of two legs of a right triangle constructed with the legs parallel to the graph axes and with the graph line along the hypotenuse. Figure 4 illustrates this, the slope being $\Delta y/\Delta x$.

So far this discussion has been strictly mathematical. Now let's consider a fairly realistic physical example. Figure 5 shows the curve from measurements of the velocity of a moving body as a function of time. If we use letters v for velocity and t for time, we'd expect this curve to be described by the relation:

$$v = v_0 + at \qquad (5)$$

Here the constant a (acceleration) is the slope of the line, while v_0 plays the role of b in Eq. (3). These two constants are physically significant, and we wish to find their values from the graph.

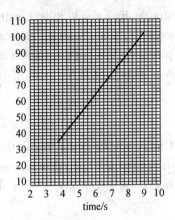

Figure 5 Measuring a slope on linear graph paper

We choose two points on the line at $t = 4.25$ and 8.75, with corresponding values of velocity: 42 cm/s and 98 cm/s. Mark these points on Figure 5, to confirm these values. The slope of the line is therefore:

$$a = \frac{(98 \text{ cm/s} - 42 \text{ cm/s})}{(8.75 \text{ s} - 4.25 \text{ s})} = \frac{56 \text{ cm}}{4.5 \text{ s}^2} = 56/4.5 \text{ cm/s}^2 = 12.44 \text{ cm/s}^2 \qquad (6)$$

The lengths are expressed in the units marked on the graph axes. The calculated slope is therefore independent of the particular choice of units, of the way you choose to label the graph scale divisions, and is also independent of the size of the graph paper.

Intercepts. The values of the intercepts are often physically significant. They can be simply read from the graph. If the $x = 0$ and $y = 0$ axes happen to be within the graph's boundaries. In the equation $y = mx + b$, the intercept is b. The v intercept of Figure 5 is the value of v when $t = 0$. It has the same units and dimensions as y. If, as in this case, the v intercept does not lie within the area of the graph, it may be calculated using the slope and one value taken from a point on the fitted line. Take the point $v = 98$ cm/s when $t = 8.75$ sec.

$$v = v_0 + at, \text{ in our case, } v = v_0 + 12.44\, t$$

so,

$$v = v_0 - 12.44\, t = 98 - 12.44(8.75) = -10.89 \text{ cm/s}$$

A check of the graph, Figure 5, shows that this looks reasonable.

Straightening a Curve. When it is possible to convert an experimental relation to a straight line graph it is usually useful to do so. Look for such opportunities. For example, when studying gases at constant temperature we find that $PV = C$ where P is pressure, V is the gas volume and C is constant. The graph of P vs. V is a branch of a hyperbola. But if we graph P-$1/V$, or V-$1/P$, the data would fall on a straight line.

$$P = \left(\frac{1}{V}\right)C \qquad (7)$$

One reason for doing this is that it is easier to fit the experimental data with a ruler-drawn straight line, than to draw the best hyperbola on a P-V graph. Another advantage is that the P-$1/V$ graph has a slope.

$$C = \frac{\Delta P}{\Delta\left(\frac{1}{V}\right)}$$

Therefore the constant C is easily determined from the straight line. This constant was not evident, nor was it easy to determine from the P-V graph!

Inexpensive electronic calculators make it so easy to manipulate data that there is no good excuse to pass up an opportunity to "linearize" experimental graphs.

[Exercises]

In each case state how you could plot (x,y) data on linear paper to obtain a straight line graph. What quantity in the equation is determinable from the slope of the straight line? What quantity is determinable from an intercept?

(1) $x(y + 1) = 3$
(2) $1/x + 1/y = 5$
(3) $y = Ae^{-x}$
(4) $y = (A - x)$
(5) $y^2 + x^2 = 7$

2.5.2 Linear Relations

The simplest curve fitting problem is that of fitting a straight line to a set of data. The problem is to find the slope of the line and its x and y intercepts. Two simpler cases frequently occur.

(1) The line may be known to be horizontal and only the y intercept is required.
(2) The intercepts may be of no interest and only the slope needs to be calculated.

The methods for fitting linear relations are of great importance because nonlinear problems may often be reduced to linear ones by an appropriate change of variable. Thus the relation $Y = bx^2$ may be converted to the linear relation $Y = bQ$ by letting $Q = x^2$ or it could be converted to $\log Y = 2 \log x + \log b$, and then to $A = 2B + C$ by letting $A = \log Y, B = \log x$ and $C = \log b$.

(1) **Horizontal Line.** This case requires only a simple average of the data values. The average of N different quantities Q_i is

$$\langle Q \rangle = \frac{\sum_{i=1}^{N} Q_i}{N} \tag{1}$$

This method gives equal weight to each Q_i. If the experimenter has knowledge that the individual data values are **not** equally reliable, this average will not give the best average value.

Weightings. Suppose the experimenter has independent means for assigning an uncertainty to each data value. A graph of such data would have error bars of various sizes on each the data points. In taking an average it is desirable to have the least accurate data values influence the result the least. This can be done by assigning weightings to each piece of data, giving the greatest weight to the most accurate data. Then the **weighted average** can be calculated

$$\langle Q \rangle = \frac{\sum_{i=1}^{N} Q_i W_i}{\sum_{i=1}^{N} W_i} \tag{2}$$

In an elementary treatment of errors, the weighting factor W_i might simply be chosen to be the reciprocal of the error in Q_i. But if you want to obtain an average consistent with the least squares criterion the weighting factor must be taken to be the reciprocal of the square of the standard deviation.

(2) **Slope of a straight line.** We now consider the case where only the slope of a straight line is required, there being no need to calculate the intercepts.

Successive differences: It might seem that the problem could be solved by simply calculating the *average* slope of the line, by finding the slope between adjacent pairs of data points, then averaging all such slopes.

Consider, for example, a case where the data points are equally spaced along the x-axis, with spacing L, and the y values are a, b, c, d, e, f, g, and h. The slope in the first interval is given by $(b-a)/L$, and the average slope is

$$\frac{(b-a) + (c-b) + (d-c) + (e-d) + (f-e) + (g-f) + (h-g)}{7L} = \langle m \rangle \qquad (3)$$

But notice that intermediate data points in the numerator cancel out and the equation reduces to

$$(h-a)/7L = \langle m \rangle \qquad (4)$$

Only the first and last data points contributed to the average. The result is merely the slope of the line between points a and h. This probably is not the best fit line, and the effort of taking the other data points, and "calculating" with them, is wasted. A better calculation method is needed, one in which all the data contributes to the average.

The method of differences avoids the difficulty mentioned above. To apply this method, first divide the data points into two equal groups (a,b,c,d) and (e,f,g,h). Then calculate slopes between the first points of each group, the second points, and so on, then average these slopes:

$$\langle m \rangle = \frac{\frac{(e-a)}{4L} + \frac{(f-b)}{4L} + \frac{(g-c)}{4L} + \frac{(h-d)}{4L}}{4}$$

$$= \frac{(e-a) + (f-b) + (g-c) + (h-d)}{16L} \qquad (5)$$

The intermediate readings do not cancel out of this equation.

Weighted successive differences: The successive differences method can be modified in such a way that intermediate data points do not cancel out, and the slope obtained is in fact the best fit. In effect, the method applies weightings to the successive differences, then averages them. The formula for the average slope of the line $y = mx + b$ is:

$$\langle m \rangle = \frac{\sum_{i=1}^{n-1}[(n-i)i]9(y_{i+1} - y_i)}{\frac{n}{6}(n^2 - 1)L} \qquad (6)$$

where n is the number of data points y_i, and $L = x_{i+1} - x_i$. The factor $(n-i)i$ weights the points most heavily for the intermediate values of i and least for the smallest and largest values of i. For 7 data points, the weighting factors are 6/56, 10/56, 12/56, 10/56, and 6/56. This equation applies only to the case where the interval between values of x is constant. This formula is easier to calculate than the least squares formulae, and is well suited to computer solution. It is, in fact, equivalent to and derivable from the least squares derived slope.

(3) **The method of least squares linear regression.** Legendre, in 1806, stated the principle of least squares for determining the best curve to fit a set of data. The principle asserts that the best curve is the one for which the sum of the squares of the deviations of the data points from the curve is smallest. To illustrate, suppose we have a set of n data values of y and x, such that to each x_i there is a corresponding value y_i. Furthermore, assume the errors are primarily in the y_i, so the x_i is assumed error free. If we choose a curve to approximate this data, it will not pass through each point. There will be a deviation in each point. The least squares criterion says that of all the possible curves one might choose, the "best" one is that for which $\sum_{i=1}^{n}(\Delta y_i)^2$ is smallest.

The reader may see in the formulation of this rule a hint as to why it is intimately connected with the standard deviation as a measure of error.

It might seem that the application of the rule to curve fitting would be difficult, if not impossible, for there are an infinity of possible curves to test! But one usually has a good idea in advance whether the best curve should be straight, parabolic, exponential or whatever, so all that remains is to determine its parameters. It is worth remarking that if there n parameters to determine, there must be at least n data points-preferably quite a few more than n to get a better fit.

Furthermore the methods of calculus allow the derivation of standard formulae for the parameters. We now state the formula without proof, for the straight line case.

Let the data points be (x_i, y_i) where $i = 1, 2, \cdots, n$. We want to fit a straight line $Y = mx + b$ to this data. (Upper case Y is used here, because values of Y_i obtained from the formula for the fitted curve will **not in general** be the same as the data points y_i). The slope of the line is given by

$$m = \frac{n \sum x_i y_i - \sum x_i \sum y_i}{n \sum x_i^2 - (\sum x_i)^2} \tag{7}$$

The summations being over i from 1 to n. The y intercept is given by

$$b = \frac{(\sum w_i x_i^2)(\sum w_i y_i) - (\sum w_i x_i)(\sum w_i x_i y_i)}{(\sum w_i)(\sum w_i x_i^2) - (\sum w_i x_i)^2} \tag{8}$$

Notice that the denominators need only be calculated once. The standard deviations of the slope and the intercept may also be found. The standard deviation of the y intercept is

$$S_b = S_y \sqrt{\frac{\sum x_i^2}{n \sum x_i^2 - (\sum x_i)^2}} \tag{9}$$

S_y is the standard deviation of the individual data values from the fitted line, given by

$$S_y = \sqrt{\frac{\sum (\Delta y_i)^2}{n - 2}} \tag{10}$$

Δy_i represents the deviation of y_i from the fitted line and n is the number of data points.

The standard deviation in the slope is given by

$$S_m = S_y \sqrt{\frac{n}{n \sum x_i^2 - (\sum x_i)^2}} \tag{11}$$

If the relationship between two sets of data (x and y) is linear, when the data is plotted (y versus x) the result is a straight line. This relationship is known as having a **linear correlation** and follows the equation of a straight line $y = m x + b$. Below is an example of a sample data set and the plot of a "best-fit" straight line through the data.

x	y
1.0	2.6
2.3	2.8
3.1	3.1
4.8	4.7
5.6	5.1
6.3	5.3

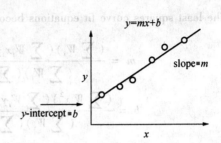

If we expect a set of data to have a linear correlation, it is not necessary for us to plot the data in order to determine the constants **m (slope)** and **b (y-intercept)** of the equation $y = mx + b$. Instead, we can apply a statistical treatment known as **linear regression** to the data and determine these constants.

Given a set of data (x_i, y_i) with n data points, the slope and y-intercept can be determined using the following:

$$m = \frac{n\sum(xy) - \sum x \sum y}{n\sum(x^2) - (\sum x)^2}; \qquad b = \frac{\sum y - m\sum x}{n}$$

(Note that the limits of the summation, which are i to n, and the summation indices on x and y have been omitted.)

It is also possible to determine the correlation coefficient, r, which gives us a measure of the reliability of the linear relationship between the x and y values. A value of $r = 1$ indicates an exact linear relationship between x and y. Values of r close to 1 indicate excellent linear reliability. If the correlation coefficient is relatively far away from 1, the predictions based on the linear relationship, $y = mx + b$, will be less reliable.

Given a set of data (x_i, y_i) with n data points, the correlation coefficient, r, can be determined by

$$r = \frac{n\sum(xy) - \sum x \sum y}{\sqrt{[n\sum(x^2) - (\sum x)^2][n\sum(y^2) - (\sum y)^2]}}$$

(4) Weighted least squares. If the data points have different standard deviations, S_i, and we define weighting factors

$$W_i = \frac{1}{(S_y)_i^2} \tag{12}$$

The least squares curve fit equations become

$$m = \frac{(\sum W_i)(\sum W_i x_i y_i) - (\sum W_i x_i)(\sum W_i y_i)}{(\sum W_i)(\sum W_i x_i^2) - (\sum W_i x_i)^2} \quad (13)$$

$$b = \frac{(\sum W_i x_i^2)(\sum W_i y_i) - (\sum W_i x_i)(\sum W_i x_i y_i)}{(\sum W_i)(\sum W_i x_i^2) - (\sum W_i x_i)^2} \quad (14)$$

【Exercises】

1. Write a compact formula for the successive differences method, using the summation symbol, and compare it with the formula for the weighted successive differences method.

2. An experiment gives the data:

X	Y
12	4.5
13	10.0
14	19.0
15	25.5
16	37.0
17	44.0
18	49.0
19	53.0
20	61.5

   Find the slope of this straight line by the method of differences. An electronic calculator or computer is very desirable for these lengthy calculations.

3. Use the least squares formulae on the data of problem 2 to find all parameters: slope, y intercept, and x intercept. The y values have a standard deviation of 0.5 units. Calculate the standard deviations of the slope and of the y intercept.

4. Write and execute a BASIC or FORTRAN or PASCAL computer program to do the calculations of problem 3. Try to make the input routine general enough so you could use the program on any size set of data you might obtain in lab.

2.5.3 Non-linear relations

The least squares principle may be extended to the problem of fitting a polynomial relation, and equations corresponding to (8) through (11) may be derived. Consult more advanced references for this. Since a very large class of relations may be approximated by polynomials, this approach has a wide utility, but in many specific cases other methods are simpler.

Another approach is to transform the relation by an appropriate change of variable, so it is in the form of a linear relation. This, in effect, straightens out the curve. This procedure is often used in graphical curve fitting, by plotting the data on special graph paper with nonlinear scales, such as log, log-log, polar or other types of graph paper. Carrying the graph analogy a bit further, note that if the original curve had error bars, they too will transform when the curve is "straightened out", and this will change the weighting factors.

Example: Exponential relation: Consider a set of data A vs. C assumed to satisfy the relation $A = BC^{(2q)}$, which can be straightened by plotting on log paper, in effect transforming the relation to

$$\log A = \log B + 2q \log C$$

This is of the form $Y = mx + b$ if we use the transformation relations

$$y = \log A \qquad b = \log B$$
$$x = \log C \qquad m = 2q$$

This can be fitted by equations (7) through (10) if c is an independent parameter of negligible error, and the error is all in the variable A. But if the standard deviations of the A_i are S_{Ai}, the standard deviations of the transformed variable Y_i will become $\log(S_{Ai})$. Then the weightings of the Y_i will be

$$W_{y_i} = \frac{1}{[\log(S_{Ai})]^2}$$

The analysis now proceeds as for a straight line fit, and values of m and b are determined.

2.5.3 Non-linear relations

The least squares principle may be extended to the problem of fitting a polynomial relation, and equations corresponding to (8) through (11) may be derived. Consult more advanced references for this. Since a very large class of relations may be approximated by polynomials, this approach has a wide utility, but in many specific cases other methods are simpler.

Another approach is to transform the relation by an appropriate change of variable, so it is in the form of a linear relation. This, in effect, straightens out the curve. This procedure is often used in graphical curve fitting, by plotting the data on special graph paper with nonlinear scales, such as log, log-log, polar or other types of graph paper. Carrying the graph analogy a bit further, note that if the original curve had error bars, they too will transform when the curve is "straightened out", and this will change the weighting factors.

Example. Exponential relation. Consider a set of data A_i, C_i assumed to satisfy the relation $A = BC^{2d}$, which can be straightened by plotting on log paper. In effect transforming the relation to

$$\log A = \log B + 2d \log C$$

This is of the form $L = mx + b$, if we use the transformation relations:

$$y = \log A \qquad b = \log B$$
$$x = \log C \qquad m = 2d$$

This can be fitted by equations (7) through (10), if c is an independent parameter of negligible error, and the error is all in the variable A. But if the standard deviations of the A are S_i, the standard deviations of the transformed variable $Y = lI$ become $\log (S_i)$. Then the weightings of the Y will be

$$W_i = \frac{1}{[\log(S_i)]^2}$$

The analysis now proceeds as for a straight line fit, and values of m and b are determined.

3 CLASSICAL EXPERIMENTS

3.1 Mechanics

3.1.1 Measurement: Length and Density of a Solid Substance

Physics is a quantitative science, relying on accurate measurements of fundamental properties such as length, mass, time, and temperature. To ensure measurements of these properties are accurate and precise, instruments such as meter sticks, Vernier calipers, micrometer calipers, triple-beam balances and laboratory thermometers are often used. Furthermore, larger range and higher precision of length can be measured with more delicate instruments as laser measurement system, light interference technology and scanning probe microscopy. It is important to understand how to use these devices properly. With any measurement tool, the student should always try to achieve the greatest accuracy the apparatus allows.

【Objective】
- To know the structure and usage of various calipers.
- To collect the measurements and to calculate the significant figures.
- To study the basic knowledge of Error-transmission and to evaluate the results by uncertainty.

【Equipment and Setup】
Vernier caliper, micrometer caliper, scale reading optical microscope.

【Principle】

1. Vernier Caliper

A vernier caliper (or vernier), shown in Figure 1, is a common tool used in laboratories and industries to accurately determine the fraction part of the least count division. The vernier is convenient when measuring the length of an object, the outer diameter of a round or cylindrical object, the inner diameter of a pipe, and the depth of a hole.

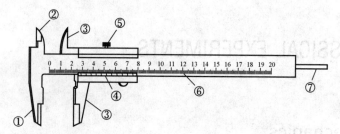

Figure 1   The vernier caliper. A common instrument
used to measure lengths, depths and diameters

(1) Parts of a vernier caliper:

①**External jaws**: used to measure external diameter or width of an object.

②**Internal jaws**: used to measure internal diameter of an object.

③**Sliding jaws**: used to move the jaws.

④**Metric Vernier scale**: gives measurements up to two decimal places (in cm).

⑤**Locking screw/Retainer**: used to block movable part.

⑥**Main scale**: gives measurements of up to one decimal place (in cm).

⑦**Depth probe**: used to measure depths of an object or a hole.

A variation to the more traditional caliper is the inclusion of a vernier scale; this makes it possible to directly obtain a more precise measurement. Vernier calipers commonly used in industry provide a precision to a hundredth of a millimetre (10 micrometres), or one thousandth of an inch. A more precise instrument used for the same purpose is the micrometer.

Vernier calipers can measure internal dimensions (using the uppermost jaws in the picture at right), external dimensions using lower jaws (in Figure 1). It depends on the manufacturer. Depth measurements can be done by using a probe that is attached to the movable head and slides along the centre of the body. This probe is slender and can get into deep grooves that may prove difficult for other measuring tools. The vernier scales may include both metric and inch measurements on the upper and lower part of the scale.

(2) Precision: The vernier consists of a main scale engraved on a fixed ruler and an auxiliary scale engraved on a moveable jaw (see part 4 and 6 in Figure 1). The moveable jaw is free to slide along the length of the fixed ruler. The main scale is calibrated in centimeters with the smallest division in millimeters. The moveable auxiliary scale has 10 divisions that cover the same distance as 9 divisions on the

main scale. Therefore, the length of the auxiliary scale is 9 mm. The difference between minima of the main scale and that of the moveable auxiliary scale is named precision of the caliper.

(3) Reading: When the vernier is closed and properly zeroed (see Figure 2), the first mark (zero) on the main scale is aligned with the first mark on the auxiliary scale. The last mark on the auxiliary scale will then coincide with the 9 mm mark on the main scale.

A reading is made by closing the jaws on the object to be measured. Make a note of where the first mark on the auxiliary scale falls on the main scale. In Figure 3, we see that the object's length is between 1.2 cm and 1.3 cm because the first auxiliary mark is between these two values on the main scale. The last digit is found by noting which line on the auxiliary scale coincides with a mark on the main scale. In our example, the last digit is 3 because the third auxiliary mark lines up with a mark on the main scale. Therefore, the length of the object is 1.23 cm.

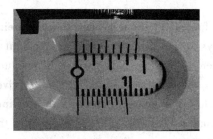

Figure 2 With the jaws closed, this is what a properly zeroed vernier caliper looks like

From http://www. clemson. edu/ces/ phoenix/tutorials/measure

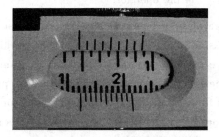

Figure 3 An example of a vernier reading. The object's length is measured to be 1.23 cm

From http://www. clemson. edu/ces/ phoenix/tutorials/measure

(4) Caution:
1) Care must be taken to insure that the vernier caliper is properly zeroed (See Figure 2).

With misuse, it is possible that the vernier will not read zero when the jaws are closed, thus leading to systematic errors. The vernier in Figure 4 is improperly zeroed. To correct this, a zero correction must be made. A correction may be either positive or negative. If the first mark on the auxiliary scale lies to the right of the main scale, then the reading is too large and the error is positive. The zero reading in

Figure 4 is + 0. 05 cm and should be subtracted from any measurement reading. Similarly, if the first mark on the auxiliary scale lies to the left of the main scale zero-mark, then the error is negative and the correction should be added from the measurement reading.

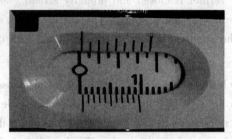

Figure 4 An improperly zeroed caliper. In this case, the error is positive (+0. 05 cm) and is to be subtracted from the measurement reading

 2) Accuracy of measurement when using a caliper is highly dependent on the skill of the operator.

Regardless of type, a caliper's jaws must be forced into contact with the part being measured. As both part and caliper are always to some extent elastic, the amount of force used affects the indication. A consistent, firm touch is correct. Too much force results in an under indication as part and tool distort; too little force gives insufficient contact and an over indication. This is a greater problem with a caliper incorporating a wheel, which lends mechanical advantage. This is especially the case with digital calipers, calipers out of adjustment, or calipers with a poor quality beam.

 3) Please put caliper back to its instrument box with outside jaws slightly apart as soon as the measurement is done.

2. Micrometer (Micrometer Screw Gauge)

A micrometer, sometimes known as a micrometer screw gauge, is a device used widely in mechanical engineering and machining for precision measurement, along with other metrological instruments such as dial calipers and vernier calipers.

Most micrometers are equipped with a ratchet (ratchet handle to far right) which allows slippage of the screw mechanism when a small and constant force is exerted on the jaw. This permits the jaw to be tightened on an object with the same amount of force each time. Care should be taken not to force the screw (particularly if the micrometer does not have a ratchet mechanism), so as not to damage the measurement object and/or the micrometer. The micrometer caliper provides for accurate measurements of small lengths and is particularly convenient in measuring the diameters of thin wires and the thickness of thin sheets.

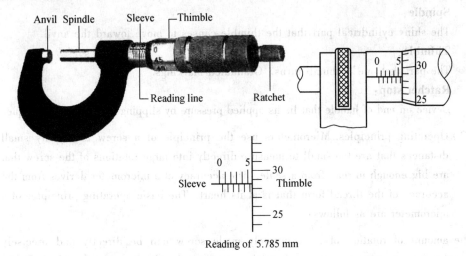

Figure 5 A micrometer caliper and an example of a micrometer reading. This particular mike has the 1.0 mm and 0.5 mm scale divisions below the reading line. In the diagram, as on some mikes, the 1.0 mm divisions are above the line and the 0.5 mm division below it. The thimble in the diagram is in its second rotation, as indicated by being past the 0.5 mm mark, so reading is 5.500 + 0.285 mm, or 5.785 mm, where the last 5 is the estimated figure (See text for details)

(1) Parts. A micrometer is composed of:

Frame:
The C-shaped body that holds the anvil and barrel in constant relation to each other. It is thick because it needs to minimize flexion, expansion, and contraction, which would distort the measurement. The frame is heavy and consequently has a high thermal mass, to prevent substantial heating up by the holding hand/fingers.

Anvil:
The shiny part that the spindle moves toward, and that the sample rests against.

Sleeve/barrel/stock:
The stationary round part with the linear scale on it. Sometimes vernier markings.

Lock nut / lock-ring/thimble lock:
The knurled part (or lever) that one can tighten to hold the spindle stationary, such as when momentarily holding a measurement.

Screw:
The heart of the micrometer, as explained under "Operating principles". It is inside the barrel.
Spindle:
The shiny cylindrical part that the thimble causes to move toward the anvil.
Thimble:
The part that one's thumb turns. Graduated markings.
Ratchet stop:
Device on end of handle that limits applied pressure by slipping at a calibrated torque.

(2) Operating principles. Micrometers use the principle of a screw to amplify small distances that are too small to measure directly into large rotations of the screw that are big enough to read from a scale. The accuracy of a micrometer derives from the accuracy of the thread form that is at its heart. The basic operating principles of a micrometer are as follows:

The amount of rotation of an accurately made screw can be directly and precisely correlated to a certain amount of axial movement (and vice versa), through the constant known as the screw's lead. A screw's lead is the distance it moves forward axially with one complete turn (360°). With an appropriate lead and major diameter of the screw, a given amount of axial movement will be amplified into the resulting circumferential movement. For example, if the lead of a screw is 1 mm, but the major diameter (here, outer diameter) is 10 mm, then the circumference of the screw is 10π, or about 31.4 mm. Therefore, an axial movement of 1 mm is amplified (magnified) to a circumferential movement of 31.4 mm. This amplification allows a small difference in the sizes of two similar measured objects to correlate to a larger difference in the position of a micrometer's thimble.

(3) Reading. In older micrometers the position of the thimble is read directly from scale markings on the thimble and shaft. A vernier scale is usually included, which allows the position to be read to a fraction of the smallest scale mark. In newer digital micrometers, an electronic readout displays the length digitally on an LCD display on the instrument.

The axial main scale on the sleeve is calibrated in millimeters and the thimble scale is calibrated in 0.01 mm (hundredths of a millimeter). The movement mechanism of the micrometer is a carefully machined screw with a pitch of 0.5 mm. The pitch of a screw, or the distance between screw threads, is the lateral linear distance the screw moves when turns through one rotation.

The axial line on the sleeve main scale severs as a reading line. Since the pitch of the screw is 0.5 mm and there are 50 divisions on the thimble, when the thimble is turned through one of its divisions, the thimble moves (and the jaws open or close) 1/50 of 0.5 mm or 0.01 mm (1/50 ×0.5 mm =0.01 mm). One complete rotation of the thimble (50 divisions) moves it through 0.5 mm, and a second rotation through another 0.5 mm, for a total of 1.0 mm or one scale division along the main scale. That is, the first rotation moves the thimble from 0.00 through 0.50 mm and the second rotation moves the thimble from 0.50 through 1.00 mm. Notice that the zero mark on the thimble is used to indicate both 0.00 mm(beginning of the first rotation) and 0.50 mm (beginning of the second rotation).

Measurements are taken by noting the position of the edge of the thimble on the main scale and the position of the reading line on the thimble scale.

For example, for the drawing in Figure 5, the mike has a reading of 5.785 mm. On the main scale is read 5.000 mm plus one 0.500 mm division (scale below reading line), giving 5.500 mm. That is, in the figure, the thimble is in the second rotation of a main-scale division. The reading on the thimble scale is 0.285 mm, where the 5 is the estimated or doubtful figure (i.e., the reading line is estimated to be midway between the 28 and the 29 marks). Some mikes have vernier scales on the sleeves to help read this last significant figure.

(4) Caution:
   1) Testing and calibration.

     A standard ordinary micrometer has readout divisions of 0.01 millimetre and a rated accuracy of +/- .001 millimetre. Both the measuring instrument and the object being measured should be at room temperature for an accurate measurement; dirt, abuse, and operator skill are the main sources of error.

   2) Torque repeatability via torque-limiting ratchets or sleeves.

     An additional feature of many micrometers is the inclusion of a torque-limiting device on the thimble-either a spring-loaded ratchet or a friction sleeve. Without this device, workers may over tighten the micrometer on the work, causing the mechanical advantage of the screw to squeeze the material or tighten the screw threads, giving an inaccurate measurement. However, with a thimble that will ratchet or friction slip at a certain torque, the micrometer will not continue to advance once sufficient resistance is encountered. This results in greater accuracy and repeatability of measurements-most especially for low-skilled or semi skilled workers, who may not have developed the light, consistent touch of askilled user.

3) Please be sure to leave a proper space between anvil and spindle after a measurement.

3. Scale Reading Optical Microscope

The optical microscope, often referred to as the "light microscope", is a type of microscope which uses visible light and a system of lenses to magnify images of small samples. By adding a cross to the eyepiece, and by fixing the objective lens on a movable broad whose movement can be measured by caliper or micrometer, the optical microscope is modified into a scale reading optical microscope. It is used in measuring property of small or uneasy-to-hold specimen, like the inner diameter of a capillary, the diameter of a small bead. The ordinary accuracy is 0.01 millimetre.

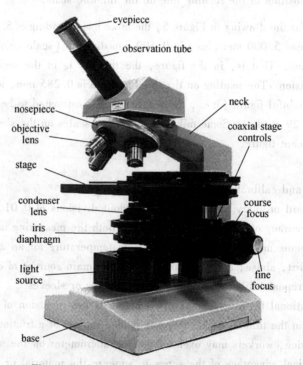

Figure 6   A scale reading optical microscope

(1) Components. A scale reading optical microscope is derived from conventional microscopes of similar basic capabilities by:

1) adding a cross to the eyepiece (to positioning a brightly lit specimen)

2) fixing the objective lens on a movable broad (whose movement can be measured by caliper or micrometer), i. e., one places the stage micrometer on the microscope stage

3) replacing the normal stage with a rotating stage (typically graduated with vernier scalesfor measuring the height of the liquid in a test tube etc.)

(2) Operation. The optical components of a modern microscope are very complex and for a microscope to work well, the whole optical path has to be very accurately set up and controlled. Despite this, the basic optical principles of a microscope are quite simple. When you view an object through a microscope, you place the object on a glass slide and cover it with the stage clips for stationary.

1) According to the characters of the specimen, the location of the scale reading optical microscope can be settled. The specimen can be put right before or down the objective lens.

2) A clear performance of the cross is made by moving the objective lens up and down or forward and backward.

3) In order to see a clear performance, the distance between the objective lens and the specimen can be changed by adjusting the course focus.

4) One thin thread of the cross can be paralleled to the main scale by rotating theeyepiece, while the other is tangent to one side of the specimen by rotating the fine focus or moving the specimen. A data X_0 corresponding to the initial location of the specimen can be read from the main scale and the thimble scale. After only moving the fine focus to tangent the other side of the specimen, another data X_1 corresponding to the final location of the specimen can be obtained. The length in observe is $L = |X_1 - X_0|$.

(3) Caution.

1) To prevent damage to the microscope, do not let the lens actually touch the object, i. e., notice the how close the high power lens comes to the stage. Use caution when handling microscope, as they can break easily and hurt you.

2) One side of the cross has to be parallel to the main scale.

3) To eliminate personal error, the direction of rotation of the micrometer drum should keep same during each measurement procedure. If the moving direction changed, turn the micrometer drum at least several full turns to ensure that the screw has meshed fully with the gear rack, be sure to squeeze the gap completely.

4. Balances

Balances are used for precision mass measurement, because unlike spring scales their accuracy is not affected by differences in the local gravity, which can vary by almost 0.5% at different locations on Earth. A change in the strength of the gravitational field caused by moving the balance will not change the measured mass, because the moments of force on either side of the balance beam are affected equally.

Very precise measurements are achieved by ensuring that the balance's fulcrum is essentially friction-free (a knife edge is the traditional solution), by attaching a pointer to the beam which amplifies any deviation from a balance position; and finally by using the lever principle, which allows fractional masses to be applied by movement of a small mass along the measuring arm of the beam, as described above. For greatest accuracy, there needs to be an allowance for the buoyancy in air, whose effect depends on the densities of the masses involved.

【Procedure】

(1) Find the mass of the solid cylinder, m, using the double pan balance. Record m and its uncertainty Δm. Remember that the uncertainty consists of two parts:
the instrumental uncertainty and the uncertainty associated with the experimenter's ability to accurately perform the measurement (here, the reading error). In this case, we have said that the age and condition of the balance has introduced an instrumental uncertainty of ± 0.5 g, and this overwhelms any reading error.

(2) Measuring the heights (h), the inside (d) and outside (D) diameters at six different locations of the hollow cylinder with a vernier caliper, and filling the data in Table 1.

(3) The volume measurements at each of the hollow cylinder were found by

$$V = \frac{\pi h(D^2 - d^2)}{4}.$$

(4) The density is given by $\rho = \frac{m}{V}$.

(5) Measuring the diameter of a steel wire by the micrometer for six times at different places, and filling the data in Table 2.

(6) Measuring the diameter of a metal wire by a scale reading optical microscope for six times at different places, and filling the data in Table 1.

3 CLASSICAL EXPERIMENTS

[Data and Analysis]

Table 1 Height, inner and outer diameter of the
hollow cylinder (measured by the vernier caliper $\Delta_B = 0.02$ mm)   (unit: mm)

Times	Height h	Inner diameter d	Outer diameter D
1			
2			
3			
4			
5			
6			
Mean value			
$S = \Delta_A$			
$\Delta = \sqrt{\Delta_A^2 + \Delta_B^2}$			

Where standard error of the mean

$$S = \sqrt{\frac{\sum(\text{Measured} - \text{Mean value})^2}{\text{total times} - 1}} = \sqrt{\frac{\sum(x_i - \bar{x})^2}{n-1}}$$

Volume and density of the hollow cylinder:

Mean volume: $\bar{V} = \dfrac{\pi \bar{h}(\bar{D}^2 - \bar{d}^2)}{4} =$

Uncertainties:

$$\frac{\Delta V}{V} = \sqrt{\left(\frac{\partial \ln V}{\partial d}\right)^2 (\Delta d)^2 + \left(\frac{\partial \ln V}{\partial D}\right)^2 (\Delta D)^2 + \left(\frac{\partial \ln V}{\partial h}\right)^3 (\Delta h)^2}$$

$$= \sqrt{\left(\frac{2d}{D^2 - d^2}\right)^2 (\Delta d)^2 + \left(\frac{2D}{D^2 - d^2}\right)^2 (\Delta D)^2 + \left(\frac{\Delta h}{h}\right)^2}$$

$$= \Delta V = \bar{V}\frac{\Delta V}{V} =$$

Measurement volume $V = \bar{V} \pm \Delta V =$

Density $\rho = m/\bar{V} =$

Table 2 Diameter of the steel wire
(measured by the micrometer, $\Delta_B = 0.004$ mm) (unit: mm)

	1	2	3	4	5	6	Mean value	S
d								

The original reading =

Uncertainties: $\Delta d = \sqrt{S^2 + \Delta_B^2} =$

$$E = \frac{|\text{Measured} - \text{Accepted}|}{\text{Accepted}} \times 100\%$$

$$= \frac{\Delta d}{d} \times 100\% =$$

Measurement: $d = \bar{d} \pm \Delta d =$

Table 3 Diameter of the steel wire (measured by the scale
reading optical microscope $\Delta_B = 0.004$ mm) (unit: mm)

	1	2	3	4	5	6
Left reading d_1						
Right reading d_2						
Diameter = $|d_1 - d_2|$						

Mean diameter: $\bar{d} =$

Standard error: $S =$

Uncertainties: $\Delta d = \sqrt{S^2 + \Delta_B^2} =$

$$E = \frac{|\text{Measured} - \text{Accepted}|}{\text{Accepted}} \times 100\% = \frac{\Delta d}{d} \times 100\% =$$

Measurement: $d = \bar{d} \pm \Delta d =$

[Questions]

1. With sketches, explain in detail the operation and the reading of vernier and micrometer calipers.
2. What are the units of density in the English system of measurement?
3. Discuss at some length the practical applications to technology and industry of the methods of precision measurement you have studied in this experiment.

3.1.2 Collisions:Conservation of Linear Momentum

In mechanics experiments, the frictional force will bring some troubles such as large system error or experimental difficulties.

【Objective】

- To measure the momentum and kinetic energy of two objects before and after one dimensional collision.
- To observe that conservation of momentum is independent of conservation of energy, that is, that the total momentum remains constant in both elastic and inelastic collisions while total kinetic energy does not remain constant in inelastic collisions.
- To try to account for any change in kinetic energy in the nearly elastic collision.
- To calculate the percentage of kinetic energy which will be lost (converted to other forms of energy, notably heat) in a completely inelastic collision between an initially stationary mass and an initially moving mass; and to compare this calculation with the result of the elastic collision.

【Equipment】

A one-dimensional air track with blower, two gliders/carts, a photogate timing circuit and an analytical balance.

【Principle】

1. Test of Momentum Conservation

One of the fundamental laws of physics is that the **total momentum** of any system of particles is **conserved** (constant), provided that the net external force on the system is zero. Momentum is a vector, making its direction a necessary part of the data. For the one dimensional case the direction can only be the $+x$ direction or the $-x$ direction. For a system of more than one particle, the total linear momentum is the vector sum of the individual momenta.

Assume we have two particles with masses m_1, m_2 and initial speeds v_{10} and v_{20} which collide, without any external force, resulting in final speeds of v_{1f} and v_{2f} after the collision. Conservation of momentum then states that the total momentum before the collision is equal to the total momentum after the collision:

$$m_1 v_{10} + m_2 v_{20} = m_1 v_{1f} + m_2 v_{2f} \tag{1}$$

In a given system, the total energy is generally the sum of several different forms of energy.

Kinetic energy is the form associated with motion. In contrast to momentum, kinetic energy is **not** a vector; for a system of more than one particle the total kinetic energy is simply the sum of the individual kinetic energies of each particle. Another fundamental law of physics is that the **total energy** of a system is always **conserved**. However within a given system one form of energy may be converted to another, such as in the freely-falling body lab where potential energy was converted to kinetic energy. Therefore, kinetic energy alone is not always conserved.

(1) **Elastic Collision.** In an **elastic collision**, two or more bodies come together, collide, and then move apart again with **no loss in kinetic energy**. An example would be two identical "superballs", colliding and then rebounding off each other with the same speeds they had before the collision. Given the above example conservation of kinetic energy then apply.

$$\frac{1}{2}m_1 v_{10}^2 + \frac{1}{2}m_2 v_{20}^2 = \frac{1}{2}m_1 v_{1f}^2 + \frac{1}{2}m_2 v_{2f}^2 \qquad (2)$$

Solving these simultaneous equations we get:

$$v_{1f} = \frac{(m_1 - m_2)v_{10} + 2m_2 v_{20}}{m_1 + m_2} \qquad v_{2f} = \frac{(m_2 - m_1)v_{20} + 2m_1 v_{10}}{m_1 + m_2}$$

or $v_{1f} = v_{10}, v_{2f} = v_{20}$

The latter is the trivial solution, corresponding to the case that no collision has taken place. Given $v_{20} = 0$, the relationship of initial and final speeds of the collision is

$$v_{1f} = \frac{m_1 - m_2}{m_1 + m_2} v_{10} \qquad v_{2f} = \frac{2m_1}{m_1 + m_2} v_{10} \qquad (3)$$

(2) **Completely Inelastic Collision.** In an inelastic collision, the bodies collide and come apart again, but **some kinetic energy is lost**. That is, some kinetic energy is converted to some other form of energy. An example would be the collision between a baseball and a bat. If the bodies collide and stick together, the collision is called **completely inelastic collision**. In this case, **much of the kinetic energy is lost in** the collision. That is, much of the kinetic energy is converted to other forms of energy.

Given the final sticking velocity $v_{1f} = v_{2f} = v$, there is

$$m_1 v_{10} + m_2 v_{20} = (m_1 + m_2) v \qquad (4)$$

Taken $v_{20} = 0$, there comes

$$v = \frac{m_1}{m_1 + m_2} v_{10} \qquad (5)$$

(3) **Coefficient of Restitution.** The coefficient of restitution (e) of two bodies for "head-on" collision is a constant and is equal to the ratio of velocity of separation and velocity of approach. This coefficient, proposed by Newton, in the usual interpretation is independent of the energy and shapes of the colliding bodies, and is completely determined by the properties of the material from which they are made. In one-dimensional collision, since motions are along a straight line, we can use scalar representation of velocity with appropriate sign convention with respect to reference direction. Velocity in an arbitrary direction is labeled "positive" and the opposite direction "negative". The coefficient of restitution is given by:

$$e = \frac{v_{2f} - v_{1f}}{v_{10} - v_{20}} \qquad (6)$$

The coefficient of restitution (e) is a positive constant, whose value falls in the range $0 \leq e \leq 1$. This fact is very helpful in working problems based on coefficient of restitution. If this ratio evaluates to negative values, then we should be certain that there is something amiss in assigning signs of terms involved in the ratio.

If we look closely at the above ratio, then it is very easy to understand following aspects of this quantity:

1) We know that velocity of separation and approach are equal in the case of elastic collision. The value of "e" is 1 in this case.
2) On the other hand, velocity of separation is zero (0) for completely inelastic (also called plastic) collision. The value of "e" is 0 in this case. Thus, elastic and plastic collisions represent the bounding values of the coefficient of restitution (e).
3) The value of "e" falls between these bounding values for other inelastic collision. The coefficient of restitution is a fraction (final kinetic energy of the colliding system cannot be greater than initial) for inelastic collision.

(4) **Loss of Kinetic Energy During Collision.** The coefficient of the loss of kinetic energy during collision can be implied by:

$$R = \frac{\frac{1}{2}m_1 v_{1f}^2 + \frac{1}{2}m_2 v_{2f}^2}{\frac{1}{2}m_1 v_{10}^2 + \frac{1}{2}m_2 v_{20}^2} \tag{7}$$

The coefficients R and e are connected by:

$$R = \frac{m_1 + m_2 e^2}{m_1 + m_2} \tag{8}$$

From equation (8), it is proved that the kinetic energy of the objects is conserved only with the case of an elastic collision, i. e., $e = 1$. Thus, the coefficient of restitution e can be used for calculating the loss of the kinetic energy during the collision.

2. Measurement of Instant Speed

We use two photogates for instant speeds. Each of them allows to measure the time it takes the cart to go through it. The velocities are calculated by dividing the length of the plate on the cart by the measured time (speed = length/time).

〖Procedure〗

(1) Before you begin this experiment, you have to **make sure the air track is level**. First, turn the air supply on. Place a glider in the middle of the track with no initial velocity. Adjust the leveling screws until the glider remains in its initial position, not accelerating in either direction. The glider may oscillate slightly about its position. This movement is caused by air currents from the air holes in the track and should be considered normal.

(2) In the following experiments you will be dealing with nearly elastic collision in which kinetic energy and total momentum is conserved. Two cases in consideration:

1) Two carts of same masses, with one initially at rest: $m_1 = m_2, v_{20} = 0$

   a. In this experiment the mass m_1 and m_2 will be measured using an electronic balance.

   b. Keep the photogate to be "$S2$" mode.

   c. Figure 1 illustrates the experimental method used for observation of elastic collisions. In this part of the experiment, you'll observe the momenta (plural of momentum!) of a pair of gliders before and after an elastic collision. Keep the photo gates in the same positions as in the first part of the experiment. Remove the wax-filled target and needle from the gliders. Attach rubber bumpers to Gliders 1 and 2, then position Glider 2 at rest between Photo gate 1 and Photo gate 2. Both Gliders 1 and 2 will be equipped with vertically positioned-measurement flags.

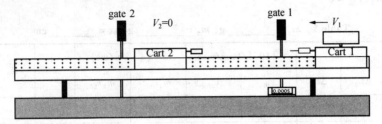

Figure 1   The initial state of the carts before collision

d. Make sure the timer is still set to gate mode with memory on. Make sure that Glider 2 (the one that is going to be hit) is placed between the two photo gates. Glider 1 should be outside the photo gates (see Figure 1). The first time measurement ("time before") will be made by giving Glider 1 a push. Push it gently (we'll explain why in a moment). As Glider 1 passes through Photo gate 1, a time interval will be measured. The initial velocity, momentum and kinetic energy of Glider 1 can be computed from the time measured, and from the mass of Glider 1. Now, this is why you want to push Glider 1 gently. You want it to hit Glider 2 so that Glider 2 will start moving, but Glider 1 will stop moving. Basically you are transferring all the kinetic energy of Glider 1 to Glider 2.

e. The second velocity you will measure is the velocity of Glider 2 as it passes through the second photo gate. This is our **time after**. The velocity, momentum and kinetic energy of Glider 2 can be computed from the time measured, and from the mass of Glider 2.

f. Repeat this experiment three times, and record your data in the table 1 on the next page, and calculate the momentum and kinetic energy for the system before and after the collision.

2) Two different masses carts, with one initially at rest: $m_1 > m_2$, $v_{20} = 0$

a. The mass of the cart 1 can be varied by adding two small metal disks symmetrically to it and measured by an electronic balance.
b. Follow the steps described in procedure 1).
c. Repeat this experiment three times, and record your data in the table 2 on the next page, and calculate the momentum and kinetic energy for the system before and after the collision.

[Data and Calculation]

1. $m_1 = m_2, v_{20} = 0$

Table 1 $m_1 = $ _____ g; $m_2 = $ _____ g; $\Delta x = $ _____ cm

No.	t/ms		$v \times 10^{-2}$/(cm/ms)		mv/(g·cm/ms)		e	R	E%
	Δt_{10}	Δt_{2f}	v_{10}	v_{2f}	$m_1 v_{10}$	$m_2 v_{2f}$			
1									
2									
3									

This distance Δx, which you need in order to calculate the velocity, is the width of the Measurement flag.

The uncertainty $E = |$final momentum − initial momentum$|$/initial momentum $\times 100\%$

2. $m_1 > m_2, v_{20} = 0$

Table 2 $m_1 = $ _____ g; $m_2 = $ _____ g; $\Delta x = $ _____ cm

No.	t/ms			$v \times 10^{-2}$/(cm/ms)			mv/(g cm/ms)		e	R	E%
	Δt_{10}	Δt_{2f}	Δt_{1f}	v_{10}	v_{2f}	v_{1f}	$m_1 v_{10}$	$m_1 v_{1f} + m_2 v_{2f}$			
1											
2											
3											

[Caution]

(1) Don't allow the gliders to crash in to the air track stops.
(2) Set the power of the air blower to maximum because this helps to minimize the friction.
(3) Please make sure to place the glider on the air track only after the air blower has been turned on and take the glider off before the air blower has been turned off.
(4) Adjust the height of the photogate so the picket fence passes easily though it. And adjust the position of the pulley so the string can be level with the air track.
(5) Make sure to plug the photogates into the correct jacks.
(6) The glider should pass completely through photogate A before the collision.

[Questions]

1. For each run of your elastic collision experiment, you should calculate the percent difference between the initial momentum and the final momentum. Does your data indicate conservation of momentum? Is the "before" velocity of Glider 1 v_{10} equal to the "after" velocity of Glider 2 v_{2f}? Why or why not?

2. In your elastic collision experiment, you should calculate the percent difference between the initial energy and the final energy. Does your data indicate conservation of energy?

3. List some possible source of error in this part of the experiment. Are these sources of error random or systematic?

4. Comment on the uncertainties of the measurement of the total momenta and kinetic energies. HINT: If the errors in the measurement of the difference of momentum and the difference of kinetic energy is large compared to the difference of momentum and the difference of kinetic energy themselves, which would mean that our measurement instruments are not precise enough to pinpoint the change in the total momentum and kinetic energy. Is this the situation with your measurements?

[Questions]

1. For each run of your elastic collision experiment, you should calculate the percent difference between the initial momentum and the final momentum. Does your data indicate conservation of momentum? Is the "before" velocity of Glider 1 v_0 equal to the "after" velocity of Glider 2 v_{2f}? Why or why not?

2. In your elastic collision experiment, you should calculate the percent difference between the initial energy and the final energy. Does your data indicate conservation of energy?

3. List some possible sources of error in this part of the experiment. Are these sources of error random or systematic?

4. Compute the uncertainties of the measurement of the total momenta and kinetic energies. HINT: if the errors in the measurement of the difference of momentum and the difference of kinetic energy is large compared to the difference of momentum and the difference of kinetic energy themselves, which would mean that our measurement instruments are not precise enough to pinpoint the change in the total momentum and kinetic energy. Is this the situation with your measurements?

3.1.3 Measurement of Rotational Inertia of Rigid Body

〖Objective〗
- Master the methods of measurement the rotational inertia of rigid body.
- Understand how to process data by graphing methods. Study linear fitting and curve straightening.
- Observe the relationship between the rotational inertia of rigid body and distribution of mass.

〖Equipment〗
Rotational inertia of rigid body apparatus; universal computerized millisecond meter; aluminous circle; aluminous plate; traction poise.

〖Principle〗

1. The Empty Rotational System

The empty rotational system consists of a stage and a wheel. I_0 is rotational inertia of the empty rotational system to axis when it rotates. The objects needed to be measured are the aluminous circle and aluminous plate. The objects are placed on the stage to get the rotational inertia of them vs. central axis when the rotational inertia of the whole rotational system is I ($I = I_0 + I_x$). I_x is calculated by the superposition of rotational inertia,

$$I_x = I - I_0 \tag{1}$$

There are two external moments acted on the whole rotational system (see Figure 1,2). One is tensile moment of string ($M = Fr$), r is the radius of the wheel with string rounded; the other is frictional moment from axis M_μ. $m_1 g - F = m_1 a$ is got by the second law of Newton when the weight is falling down. m_1 is the mass of weight together with poise hook; a is the acceleration of the falling weight; at normal situations, $a \ll g$, so it can be thought approximately that $F = m_1 g$, the formula below can be deduced by rotational law,

$$\begin{cases} m_1 gr - M_\mu = I\beta & (\text{rotation}, \beta \text{ is a positive constant}) \\ -M_\mu = I\beta' & (\text{rotation}, \beta' \text{ is a negative constant}) \end{cases} \tag{2}$$

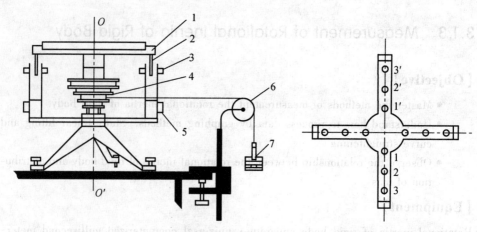

Figure 1 Rotational inertia of rigid body apparatus
1. stage 2. block stick 3. light-electric door
4. wheel 5. base 6. pulley 7. weight

Figure 2 Look down picture of stage

So,

$$M_\mu = \frac{-\beta'}{\beta - \beta'} m_1 gr; \quad I = \frac{m_1 gr}{\beta - \beta'} \tag{3}$$

Note: here β' is negative; I is moment of inertia. The moment of inertia of the empty rotational system I_0 can be measured by the same methods. Then I_x, the moment of inertia of objects can be calculated from, $I_x = I - I_0$. The key points of measurement of moment of inertia I are how to define angular acceleration β, fictional moment from axis M_μ, and β', as can be seen from Eq. (3).

Frictional moment of inertia is rarely affected by the speed of rotation when the system rotates. So it is believed to be a constant moment. Thus the rotation can be seen as a motion with constant angular acceleration. So,

$$\theta = \omega_0 t + \frac{1}{2}\beta t^2 \tag{4}$$

θ_1, θ_2 is measured as angular displacement of the rotation system turned from a same start point to two different angles, the time needed for the travel to θ_1, θ_2 is record as t_1, t_2. From Eq. (4), the following is got,

$$\begin{cases} \theta_1 = \omega_0 t_1 + \frac{1}{2}\beta t_1^2 \\ \theta_2 = \omega_0 t_2 + \frac{1}{2}\beta t_2^2 \end{cases} \tag{5}$$

where ω_0 is the angular speed from the start point, θ_1, θ_2 is the angular displacement corresponding to t_1, t_2. The parameters were set as $\theta_1 = 2\pi, \theta_2 = 8\pi$ here.

The angular acceleration of the system when it rotates is deduced by the two equations from Eq. (5),

$$\beta = \frac{2(\theta_1 t_2 - \theta_2 t_1)}{t_1^2 t_2 - t_2^2 t_1} \tag{6}$$

The stage undergoes equally decelerating rotational motion by the frictional moment when the weight on the end of the string reached ground which resulted in disappearance of the tensile moment. The angular acceleration can be deduced by same methods.

$$\beta' = \frac{2(\theta_1 t_2' - \theta_2 t_1')}{t_1'^2 t_2' - t_2'^2 t_1'} \tag{7}$$

t_1, t_2, t'_1, t'_2, from records by the millisecond meter was put into formula Eq. (7) and β, β', can be deduced there. Moment of inertia I and the frictional moment M_μ can be got from Eq. (3).

2. Graphing methods

If the angular speed from start point is $\omega_0 = 0$, from Eq. (4),

$$\beta = \frac{2\theta}{t^2}$$

when it's applied to the first formula of Eq. (2),

$$m_1 gr - M_\mu = \frac{2I\theta}{t^2}$$

that is,

$$m_1 = \frac{2I\theta}{gr} \cdot \frac{1}{t^2} + \frac{M_\mu}{gr} = K\frac{1}{t^2} + m_\mu \tag{8}$$

If θ, r is defined, M_μ is constant, m_1 should have a linear relationship with $1/t^2$. A series of m_1 vs. t values can be got by recording t value of the system passing the same angle while the alteration of the poise weight m_1 value was applied. Thus the moment of inertia I and frictional moment M_μ can be defined by graphing methods.

3. Introduction of TH-4

The universal computerized millisecond meter (see Figure 3) was designed for the measurement of solid moment of inertia. The microcomputer (SCM) was the central part of it and the memory function was created with the maximum recall of 99 sets of measured records. The records desired can be output at any moment. People can select one of the measure methods based on the requirement from several methods. The range of measurement was 0 ~ 99.999 9 seconds with a scale of 0.1 millisecond.

The face panel of this machine: 8 - digit display, the left two digits is for recording the round of counting; the right six digit is for recording the time. In the mini-keyboard, "0 - 9" is the number key; " - " is minus key; "β" is the angular acceleration key; "send" is the key to send the data to computer (not activated yet); "set/up" is the key to sort the data incrementally; "set/down" is the key to sort the data descending. "Reset" is the key to reset the machine when it doesn't work well and erase all the settings and data restored.

The back panel: two of channels for counting; output power of 2.2 V at the same time.

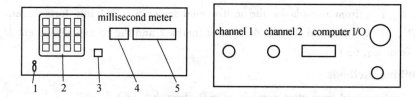

Figure 3 Universal computerized millisecond meter
(Left, face panel; Right, back panel)
1. Switch 2. Mini keyboard 3. Reset 4. Display of counting times 5. Display of travel time

Operation instruction:

(1) Power on. The switch is on the left bottom corner of the face panel. The 8 - digit display shows "88 - 888888", otherwise press the "reset" key.

(2) Press "set" key, the display shows "P - 0199", "01" means one block of the light to count once, and "99" means the maximum counting is 99 times. If setting the machine as "0129", means one block of light to count once and the maximum counting is 29 times. If setting the machine as "0211", means two blocks of the light to count once and the maximum counting is 11 times. After setting the machine as desired, one more push of the "set" key is needed to go back, then the system undergoes into counting state with the display says "88 - 888888", the machine is ready to use now.

(3) Output t values, all the data is stored in the SCM after one time measurement. In order to get the data for analysis from the SCM, manually extraction of the data is required according to the suggestions below.

3 CLASSICAL EXPERIMENTS

If the data of counting times under 10 is needed, directly input the number of counting times (e. g. if you need the t value of the seven times counting, input 7 through number keyboard); if the counting times number is over 10, then continuously input the number you need with interval shorter than 2 second, otherwise the machine would mistake the number as two times input as both smaller than 10. For example, if the 23rd times counting is needed, then first input 2, and quickly input 3, the 23rd results t value would be displayed.

(4) Output the data ascendingly or descendingly; "up" or "down" is the key to use here.

(5) Subtraction of two t values: for example, "23" - "21", first input "2", "3", second input " - ", then the display would say "P - ", third input "2", "1" (no displaying of "21"), fourth get the result. Note if the subtraction is a number smaller than 10, a moment will be needed to wait to get the result. "P - " is only the absolute values, no minus or plus is involved.

(6) Show "β" value: Directly calculation of "β" value is provided by this universal computerized millisecond meter. The SCM already set the angular displacement θ_1, θ_2 as $2\pi, 8\pi$ ($\theta_1 = 2\pi, \theta_2 = 8\pi$), corresponding to time of the 3rd and 9th counting. Just press one time to get "β" value displayed as "1$\cdots$" after one measurement. If press "β" key twice continuously, the display shows "2 - ", this is the minus angular acceleration under the function of frictional moment β'. The minus or plus symbol needs to be noticed.

(7) Push "counting" key to restart measurement. Do not push "reset" key, otherwise the setting will be erased.

[Procedure]

Note: Something needs your attention when you do this experiment:
(1) The length of string is about 120 cm. Do not use a string much longer than this;
(2) The string should round the wheel side by side. Do not cross over with each other;
(3) One end of string is at the same height as the top of wheel, and the string is set to be along with the pulley, not twist with the pulley;
(4) Make sure that the weight falls down from zero velocity;

1. Measure the Moment of Inertia of Aluminous Ring to Center Axis

(1) First put the aluminous ring on the stage and measure the whole moment of inertia of system I. Two angular displacement θ_1, θ_2 are considered to be 2π, 8π respectively. Millisecond meter is set to be "0129". m is the mass of four poises together with the hook. r equals 25 mm. When setting the string onto the appara-

tus, first make a loose knot at one end of string at which there's no poise hanging. Second inset the knot end of string into the indentation of the edge of string wheel with radius of 25 mm (the third layer from top). Third rotate the system to get the string onto the string wheel axis desired via the pulley. Hold the stage with one hand and press the key "counting" on millisecond meter with the other hand. Then remove the hand on the stage and the system will rotate from zero velocity under the external moment (M) and frictional moment (M_μ). The weight will reach the floor after 9th time counting. Because of M_μ, the system continues to rotate after the weight reached floor till the millisecond meter stop counting.

Note: t values: t_1 (the 3rd time), t_2 (the 9th time), the angular displacement is 2π, 8π respectively.

t'_1 ($23^{rd} - 21^{st}$ time), t'_2 ($29^{th} - 21^{st}$ time), the angular displacement is also 2π, 8π respectively.

Press "β" key to get β value when display shows "1⋯"

Press "β" key to get β' value when display shows "2⋯" β' value is minus here.

Record "β" value and "β'" value.

Repeat the procedure above to 6 times, 6 sets of data should be got here.

Scale and record the mass, inner radius, outer radius of aluminous ring.

(2) Remove the aluminous ring from the stage and measure the moment of inertia of the stage I_0 (the empty rotation system). I_0 is deduced the same way as I mentioned above in (1).

(3) Calculate the moment of inertia of aluminous ring vs. center axis I_x according to Eq. (1).

(4) Compare the I_x value above with the value from theoretical formula deduction.

$$I_{theory} = \frac{1}{2} m_2 (R_{inner}^2 + R_{outer}^2)$$

m_2 is the mass of aluminous ring, R_{inner}, R_{outer} is the inner radius, outer radius of aluminous ring respectively.

2. Processing the Data by Graphing Methods

Measure the moment of inertia of aluminous plate vs. center axis.

(1) Measure the moment of inertia of the whole system I. Place the aluminous plate onto the stage. The angular displacement is set to be 8π, the radius of string wheel is 25 mm. m_1 is the mass of hook together with weight. m_1 has 8 different values because the poise mass is added piece by piece each time starting from one piece.

The t values are recorded by the millisecond meter when the system rotates 8π with a mass of m_1 each time. To make sure that $\omega_0 = 0$, counting must begin at exact moment that the system starts rotate from zero velocity. In order to do this, the block stick must be put in the right place. Face the working light-electric door when operating the apparatus, set the edge of the shade of block stick tangent with the edge of hole of coming light to satisfy the condition of $\omega_0 = 0$. According to the principle of the experiment, linear relationship is found between m_1 and $1/t^2$. The slope $K = \Delta m_1/\Delta(1/t^2) = 2I\theta/gr$ can be calculated by graphing method, so the moment of inertia $I = Kgr/2\theta$ can be deduced.

(2) The moment of inertia of the stage (empty system) I_0 can be measured from procedure 1.

(3) The moment of inertia of aluminous plate I_x can be deduced from formula Eq. (1).

(4) Calculate the moment of inertia of aluminous plate vs. center axis I_{theory} by theoretical formula, and compare it with the value got from experimental measurement.

【Data and Analysis】

1. Measure the Moment of Inertia vs. Center Axis of Aluminous Ring

The parameters in this experiment:

The mass of one piece of weight, 5.0 g; the mass of hook, 2.5 g.

The radius of aluminous ring: $R_{outer} = 120.00(0.02)$ mm, $R_{inner} = 105.00(0.02)$ mm

The mass of aluminous ring: $m_2 = $ _____.

The mass of weight and hook: $m_1 = 22.5(0.5)$ g.

The radius of wheel: $r = 25.00(0.02)$ mm, angular displacement: $\theta_1 = 2\pi$, $\theta_2 = 8\pi$.

(1) Experiment data of moment of inertia with luminous ring.

	1	2	3	4	5	6	Average
$\beta/(\text{rad/s}^2)$							
$\beta'/(\text{rad/s}^2)$							

The moment of inertia $I = $ _____ kg·m² $M_\mu = $ _____ N·m

(2) Experiment data of moment of inertia without luminous ring.

	1	2	3	4	5	6	Average
$\beta/(\text{rad/s}^2)$							
$\beta'/(\text{rad/s}^2)$							

The moment of inertia $I = $ _____ kg·m² $M_\mu = $ _____ N·m

(3) The moment of inertia of aluminous ring.

Value from experiment measurement:

$$I_x = I - I_0 = \underline{\qquad} \text{ kg} \cdot \text{m}^2$$

Value from theoretical deduction:

$$I_{theory} = m_2(R_{inner}^2 + R_{outer}^2)/2 = \underline{\qquad} \text{ kg} \cdot \text{m}^2$$

$$E = \frac{I_{theory} - I_x}{I_{theory}} \times 100\% = \underline{\qquad}$$

2. Measurement of Moment of Inertia of Aluminous Plate vs Center Axis (Processing to Data by Graphing Method)

Parameters in this assay:

Radius of aluminous plate: $R = 120.00(0.02)$ mm; Radius of wheel: $r = 25.00(0.02)$ mm

Original angular velocity: $\omega_0 = 0$ 　　　Angular displacement: $\theta = 8\pi$

Mass of aluminous plate: $m_3 = \underline{\qquad}$

(1) Experiment data of moment of inertia with aluminous plate.

m_1/g						
t/s						
$(1/t^2)/\text{s}^{-2}$						

Make graph of $m_1 - 1/t^2$ on plotting paper.

Calculate: about the linear equation,

　　　Slope $K = \underline{\qquad}$;

Moment of inertia $I = Kgr/2\theta = \underline{\qquad}$.

(2) The moment of inertia of aluminous plate.

Value from experiment measurement:

$$I_x = I - I_0 = \underline{\qquad} \text{ kg} \cdot \text{m}^2$$

Value from theoretical deduction:

$$I_{theory} = m_3R^2/2 = \underline{\qquad} \text{ kg} \cdot \text{m}^2$$

$$E = \frac{I_{theory} - I_x}{I_{theory}} \times 100\% = \underline{\qquad}$$

[Questions]

What determine the rotational inertia of a rigid body?

3.1.4 The Determination of Liquid Surface Tension Coefficient

The value of the surface tension can be described by the surface tension coefficient. In this experiment, the surface tension coefficient is measured by Pulled-off method where the force is measured by force sensor.

【Objective】

- Learn to determine the liquid surface tension coefficient.
- Find out the structure of the Jonly scales and learn to handle it.
- Perceive the phenomenon of liquid surface in deeper ways.

【Equipment】

Jonly scale, beaker, weights.

【Principle】

Surface tension is a phenomenon in which the surface of a liquid, where the liquid is in contact with gas, acts like a thin elastic sheet. This term is typically used only when the liquid surface is in contact with gas (such as the air). If the surface is between two liquids (such as water and oil), it is called "interface tension".

Various intermolecular forces, such as Van der Waals forces, draw the liquid particles together. Along the surface, the particles are pulled toward the rest of the liquid, as shown in Figure 1.

Surface tension (denoted with the Greek variable γ) is defined as the ratio of the surface force F to the length d along which the force acts: $\gamma = F/d$.

Surface tension is measured in SI units of N/m (Newton per meter).

Figure 1 **The forces acting on a liquid that cause surface tension**
From http://soft-matter.seas.harvard.edu/

Case 1: To consider the pressure inside the soap bubble, we consider the radius R of the bubble and also the surface tension, γ, of the liquid. Consider a cross-section through the center of the bubble, ignoring the very slight difference in inner and outer radii, we know the circumference will be $2\pi R$. Each inner and outer surface will have a pressure of γ along the entire length, so the total. The total force from the surface tension (from both the inner and outer film) is, therefore, $2\gamma(2\pi R)$. Inside the bubble, however, we have a pressure p which is acting over the entire cross-section πR^2, resulting in a total force of $p(\pi R^2)$. Since the bubble is stable, the sum of these forces must be zero so we get:

$$2\gamma(2\pi R) = p(\pi R^2) \quad \text{or} \quad p = 4\gamma/R$$

Obviously, this was a simplified analysis where the pressure outside the bubble was 0, but this is easily expanded to obtain the difference between the interior pressure p and the exterior pressure p_e:

$$p - p_e = 4\gamma/R$$

Case 2: Analyzing a drop of liquid, as opposed to a soap bubble, is simpler. Instead of two surfaces, there is only the exterior surface to consider, so a factor of 2 drops out of the earlier equation to yield:

$$p - p_e = 2\gamma/R$$

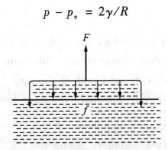

Figure 2 The forces acting on a liquid that cause surface tension

Case 3: The moment that the liquid separates from the U-shape wire, the pull F and the surface tension f will approach an equilibrium (Figure 2). So $F = 2f = 2\gamma l$, where l is the length of the U-shape wire.

Case 4: The moment that the liquid separates from the ring-shape wire, the pull F and the surface tension f will approach an equilibrium (Figure 3). So $F = \gamma\pi(D_1 + D_2)$, where D_1, D_2 are the inner and outer diameters of the ring.

【Procedure】

(1) Following figure 3, install the instrument and adjust it.

(2) Place the weight tray on the pothook of the force detector, adjust the reading of the detector U to zero. The reading U of the force detector is linear proportional to the external force F, i.e., $F = U/k$. Add weight on the tray each time by 500 mg and stop at 3 500 mg. Fill the reading U from the force detector to the table1. Draw a U-m plot and solve the slope C. Calculate k according to $k = C/g$.

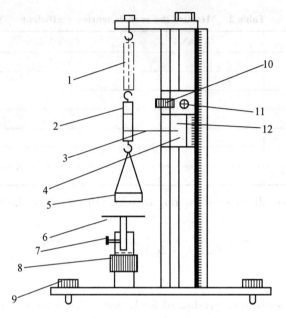

Figure 3 Instrument set up
1. spring   2. counterweight cylinder   3. pointer   4. caliper
5. scale pan   6. objective table   7\9\11. screws
8\10. vernier adjustment knob   12. mirror

(3) Measure the surface tension coefficient by Pulled-off method. Place the ring on the pothook of the force detector. Fill water into the breaker. Dip ring into the water and draw it out of the water gently and slowly. Observe the reading of the detector at the moment that the ring is out of the water. The moment that the ring-shape wire be pulled-off the liquid, reading from the detector before (U_1) and after (U_2) the separation gives out the pull F by $F = (U_1 - U_2)/k$. Fill the reading U_1, U_2 into table 2. Repeat 6 times and then calculate the surface tension coefficient γ.

【Data and Analysis】

Table 1 Scale the force detector

Mass of each weight is 500 mg, $g = 10$ m/s^2

Mass m/g	0.500	1.000	1.500	2.000	2.500	3.000	3.500
Reading U/mV							

$C = $ _____ mV/g; $k = $ _____ V/N

Table 2 Measure the surface tension coefficient

#	U_1/mV	U_2/mV	$U_1 - U_2$/mV
1			
2			
3			
4			
5			
6			
Average			

Inner and outer diameter of the ring $D_1 = 3.310$ cm; $D_2 = 3.496$ cm.

$$\bar{\gamma} = \frac{\overline{U_1 - U_2}}{k\pi(D_1 + D_2)} = \underline{\qquad} \text{ N/m}$$

【Caution】

(1) The detector needs to be pre-heated for 15 min before experiment.
(2) Don't turbulence the liquid while working on it.
(3) Keep the liquid clean and do not touch it directly by fingers.

【Questions】

1. Will liquid surface tension coefficient change as the liquid's temperature is changed?

2. What's the physical significance of the liquid surface tension coefficient?

3. How many factors can affect the liquid surface tension coefficient? List and explain them.

3.1.5 Measuring Young's Modulus with a Stretching Method

【Background】

A point particle is the simplest case when describing a motion. As we get into more complicated motions and what causes the motion, we take into account that all objects have a volume and a shape, with the assumption that they are perfectly rigid. In the real world this is not the case and how a force is applied determines how the shape of an object changes.

This distortion is quantified by the **strain**, a measure of how much the shape changes compared to its original shape. Information of the cause of the distortion such as the angle of the force and the area of the object where it was applied is contained in the **stress**. The distorting force can be applied normal (perpendicular) to a surface (either in or out), tangential to the surface, or uniformly perpendicular across the entire surface of the object. The corresponding stresses are **tension** or **compression**, **shear**, and **hydraulic**. In each case, under certain conditions the resulting strain is proportional to the stress.

In this experiment, we investigate the change in length of a wire under a varying tension. The constant of proportionality between the tensile stress and strain is called **Young's modulus**, also known as the **elastic modulus**. (Modulus is just another word for constant.)

【Objective】

- Perform an experiment to estimate the Young's modulus.
- Investigate the elasticity of materials by showing that the stress is proportional to the strain.
- To find Young's modulus for different materials.
- Analyze and discuss the reason of error existed in the experiment.

【Equipment】

Steel wire, pulley and a mass hanger, weight, Hooks, mass pieces, meter ruler, micrometer caliper, vernier caliper, micrometer screw gauge, steel or aluminum beam.

【Principle】

1. Young's Modulus

Young's modulus is named after Thomas Young, it shows when a solid body which is deformed by external force, known by Hooke's Law, in elastic deformation, the tensile stress of material is proportional to the tensile strain of it, and it determine the proportional coefficient is Young's Modulus:

$$E \ \Delta L/L = F/A$$

Where E is elastic modulus (unit N/m^2), F is the tension of a wire and A is the cross-sectional area, ΔL is the tension of a wire and L is the initial length of the wire. If L is constant, E and A are also constant, then $\Delta L \alpha F$. So the extension is proportional to the stretching force. The equation can also be used when material is being compressed.

2. Molecular Model of Stress and Strain

To see how this linear stress-strain relationship comes about, let's look at a solid metal wire on the atomic scale. The metal atoms of the wire are bound together by electrical forces. The exact nature of these forces is complicated but we can model the solid approximately as points (the atoms) in a three-dimensional lattice, interconnected with springs. If we anchor one end of the wire and stretch it out with a known force at the other end, each little spring aligned along the length of the wire will stretch by the same amount since the same applied force is transmitted uniformly to each. Although the stretch of an individual spring may be microscopic, there are a lot of them and the sum of all of these teeny changes can produce a macroscopic change in the length of the wire.

Just like a spring, for a certain range of applied forces, the wire obeys **Hooke's law**. This means that the change in length is proportional to the applied force. Also like a spring, when the applied force is removed, the wire returns to its original length. If we go beyond that certain range, if we exceed some maximum applied force, the wire or spring is permanently deformed or it breaks. The only difference between a wire and a spring is that the magnitude of the stretch. This stretching behavior is summarized in a stress-strain diagram such as the one shown in Figure 1.

As the stress is increased between points a and b on the graph, the stress-strain relationship for the wire is linear and elastic. Between points b and c the behavior is still elastic but is no longer linear. After point c, called the **yield strength**, the material enters the plastic deformation region, which means that the stretch of the wire is permanent. E. g, if the wire is stressed to point d on the graph and the stress is slowly decreased, the stress-strain curve follows the dotted line instead of the original curve and there is a leftover strain when all stress is removed. At point e the wire reaches its

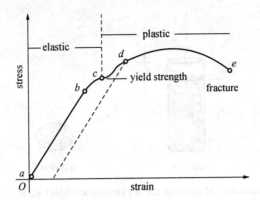

Figure 1 Typical tensile stress-strain curve for a metal. Of interest in this experiment is the linear, or elastic, region

From http://www.physics.uc.edu/~bortner/labs/physics

breaking point. Differences in the shape and limits of the stress-strain diagram determine whether a material is considered ductile or brittle, elastic or plastic.

3. Macroscopic Behavior

We assume that the material put under stress is **homogeneous** and **isotropic**. Homogeneous means that the elastic properties are the same throughout the bulk of the sample. Isotropic means that the properties do not depend on direction. Since the direction of the force is along the axis of the wire in this experiment, this means that the orientation of the material within the wire does not affect its behavior.

You are probably aware from observation of the stretching of rubber bands or the making of balloon animals that when the length of an object is stretched, its diameter decreases. This definitely happens on a smaller scale in the case of a stretched wire, but we do not take this into account in our analysis of this experiment.

The illustration in Figure 2 indicates the relevant physical quantities for defining Young's modulus:

   the applied force F
   the cross sectional area A
   the initial, unstressed length L_0
   the change in length due to the stress, ΔL

We now define the stress and strain, the **stress** S is the force per unit area:

$$S = F/A \tag{1}$$

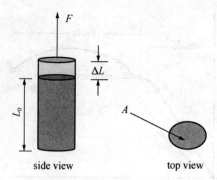

Figure 2 Quantities of interest for a cylindrical object under tensile stress
From http://www.physics.uc.edu/~bortner/labs/physics

The **strain** e is the fractional change in the length:

$$e = \Delta L / L_0 \qquad (2)$$

Within the linear, elastic region of the stress-strain diagram (from a to b on Figure 1), Young's modulus E is defined as the ratio of stress to strain:

$$E = S/e = (F/A)/(\Delta L / L_0) \qquad (3)$$

Equivalently, Young's modulus is the constant of proportionality between the stress and the resulting strain which it produces, or the slope of the line in the plot of stress vs. strain:

$$S = E \times e \qquad (4)$$

Since the strain is a ratio and therefore has no unit, Young's modulus has the same units as the stress, which is Newton per square meter (N/m^2), or Pascal(Pa).

4. Leveling the Apparatus

The equipment should be set up as shown in Figure 3. The wire is firmly held at its top end by a collet. The two kilogram mass hanger is attached to the other end, initially with no additional mass. A second collet, clamped to the wire near its lower end, connects the wire to the balancing micrometer.

There are two separate leveling procedures in this experiment. First, the entire stand must be leveled using the adjustable feet on the tripod stand. The mass holder has to be centered between the two vertical rods that support the wire. If the mass holder is making contact with either rod, adjust the feet on the tripod until the mass holder is centered.

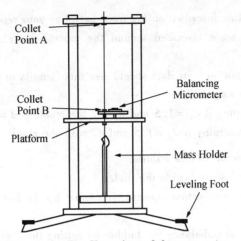

Figure 3 The Young's modulus apparatus
From http://www.physics.uc.edu/~bortner/labs/physics

The second critical procedure comes in leveling the **balancing micrometer** in order to determine the change in length of the wire. Observe the bubble level that is attached to the balancing micrometer. Turn the dial to level the micrometer, placing the bubble in an easily repeatable position. Note that the length changes are on the order of hundredths of a millimeter. Having the micrometer perfectly level, with the bubble exactly centered, is not as important as having the bubble located at a repeatable reference mark, something that is clear and easy to judge. Make sure that you always use the same vantage point when observing the bubble. Have your head and eyes in the same location every time you adjust the micrometer. Throughout the experiment, you determine the change in length of the wire by adjusting the micrometer to bring the bubble back to this original position.

The point of the micrometer screw rests in the slot of the head of a screw. This screw point must rest at the same point in the slot each time you level the micrometer.

A final thing to consider when balancing is that the micrometer assembly has some torsional freedom, which affects the bubble position. Keep the assembly twisted clockwise as far as it will go when you level the micrometer.

To record the micrometer measurement, follow this process:
(1) Using the edge of the rotating barrel as your reference mark, read the scale on the fixed barrel (up to the edge of the moving barrel) to obtain the measurement to the nearest half millimeter. Record this number as d_F.

(2) Using the center line inscribed on the fixed barrel as your reference mark, take the reading from the scale inscribed around the moving barrel. Record this number as d_R.

(3) In the Analysis (not on your data sheet) use this formula to get your measurement d in mm: $d = d_F + d_R/100$.

    1) One example: if $d_F = 1.5$ and $d_R = 7$ the result is $d = 1.57$ mm.

    2) Another example: if $d_F = 1.5$ and $d_R = 35$ the result is $d = 1.85$ mm.

Read the balancing micrometer as follows:

(1) Look at the vertical scale beside the dial.

    1) Reading from the bottom upward, measure up to but not past the top of the dial.

    2) You may have to unbalance the bubble by setting the dial at zero.

    i. Note the vertical scale reading (it may be ambiguous).

    ii. Increase (turn counterclockwise) or decrease (clockwise) the height by making one complete turn to the next zero.

    iii. Note the scale reading.

    iv. Rebalance the bubble.

(2) Record this measurement on the vertical scale as the number of millimeters.

(3) Look down at the top of the dial and record the number that coincides with the front edge of the vertical scale as the hundredths of a millimeter (i.e. if the dial is at 5, append 0.05 mm to the number of millimeters from Step (2); if the dial is at 63, append 0.63 mm).

[Procedure]

1. Procedure A. Nylon Wire and Steel wire

(1) Fix the iron supporter on the table.

(2) Tie up one side of the nylon rope to the iron supporter and the other side to a pulley with a mass hanger and fixed the pulley on the table.

(3) Make the nylon wire loose and measure the length of it with ruler.

(4) Determine the diameter of the nylon wire.

(5) Put weight on the mass hanger and record the mass add each time.

(6) Record the length of the nylon rope change each time (repeat step (5) and (6)).

(7) Change the nylon rope to iron wire and repeat the procedure (1) ~ (6) above.

2. Procedure B. Two Unknown Wire

(1) Level the stand so that the mass holder hangs freely.

3 CLASSICAL EXPERIMENTS

(2) Once the apparatus is level, use the meter stick and the vernier accessories to measure L_0, the initial length of the wire between the two collets (the length between points A and B in Figure 3).

(3) Use the Starrett micrometer to record the two numbers d_F and d_R that give the diameter d of the wire.

(4) Record as x the initial reading of the balancing micrometer with no added masses ($m = 0$) on the mass holder.

(5) Add/remove mass, measure the change in height/length. Place a 1 kg mass on the mass holder. The wire stretches and the balancing micrometer is no longer level.

(6) Turn the dial of the balancing micrometer until it is level again. (The bubble is back at the reference position. Again, be consistent in viewing this.)

(7) Record the total mass m added to the holder (do not include the mass of the holder itself).

(8) Record the new position x of the micrometer.

(9) Continue this process (steps (5) ~ (8)) for the following values of m, in this order: 1, 2, 3, 4, 5, 6, 7, 8, 8.5, 7.5, 6.5, 5.5, 4.5, 3.5, 2.5, 1.5, 0.5, and 0. That is, a. set m, b. level the micrometer, c. record x.

[Data and Analysis]

1. A. Nylon Wire and Steel wire

Uncertainty in mass m (ignore the mass of the pulley and mass hanger) ±1 g;
Uncertainty in displacement ±0.10 cm.

(1) Material type: Nylon Wire

  1) Data

Table 1   Initial Setup

L = the initial length of the wire, d = the diameter of the material.

Trail#	1	2	3	4	5	6	Avg.
L							
d							

$$\Delta = \sqrt{\Delta_A^2 + \Delta_B^2}; \quad \Delta_A = \sqrt{\sum_{i=1}^{6}(x_i - \bar{x})^2/(n-1)}$$

The measurement of L = _____, d = _____

Table 2 Tensile Force and Strain Data

Trial #	1	2	3	4	5	6	7	8	9	10	avg
Mass m(g)	0										
$L + \Delta L$ (cm)											
$L - \Delta L$ (cm)											

2) Analysis: Linear regression

First Calculate Young's Modulus according to $E = S/e = (F/A)/(\Delta L/L_0)$ = _____.

Then use linear regression method to determine the best fit line. We can assume that the equation of the best fit line is $y = a + bx$. Consider the units of m and ΔL, the gradient of the slope in fitting figure (mass v. s. extension) can be obtained as b = _____.
(Note: find two non-experimental points on the line to easily calculate the slope.)

E = _____,

Absolute error = _____,

Standard deviation in the slope = _____.

The final Young's Modulus of the nylon wire E = _____.

(2) Material type: Steel Wire

1) Data

Table 3 Initial Setup. L = the initial length of the wire,
d = the diameter of the material

Trail#	1	2	3	4	5	6	avg
L							
D							

The measurement of L = _____, D = _____

Table 4 Tensile Force and Strain Data

Trial #	1	2	3	4	5	6	7	8	9	10	avg
Mass m(g)	0										
$L + \Delta L$ (cm)											
$L - \Delta L$ (cm)											

2) Analysis: Linear regression

According to the Young's Modulus: $E = S/e = (F/A)/(\Delta L/L_0)$ = _____

Then use linear regression method to determine the best fit line. We can assume that the equation of the best fit line is $y = a + b x$. Considering the units of m and L, the gradient of the slope in fitting figure (mass v. s. extension) can be obtained as $b = $ _____. (Note: find two non-experimental points on the line to easily calculate the slope.)

$E = $ _____,

Absolute error = _____,

Standard deviation in the slope = _____.

The final Young's Modulus of the nylon wire $E = $ _____.

2. B: Two Unknown Wires

(1) Open the Young's modulus template and enter the values of g, L_0, d_F, d_R for the first wire, and the uncertainties in L, d, m, and x. (The only thing in the template, besides the functions, is the table of Young's moduli. You have to label everything else and do the plots.)

(2) Calculate d.

(3) Enter the data for m and x into two appropriately labeled columns, including the reading for $m = 0$.

(4) Calculate the strain Eq(1) and the stress in GPa Eq. (2) in adjoining columns.

(5) Propagate the error for e and S.

(6) Calculate $u\{e\}$ and $u\{S\}$ for each data point. (There is one value of $u\{e\}$ for all data points.)

(7) Graph stress vs strain, including error bars and a trendline.

(8) Calculate Young's modulus E and $u\{E\}$ in GPa using LINEST(linear regression).

(9) Compare your result to each of the values in the accompanying table of Young's moduli for different materials to identify the wire material. Assume that the error for the accepted value is the least (smallest) significant digit. If nothing matches, or you have more than one match, what is your best guess?

Table 5 Record the position x of the micrometer

m/kg	$x_i(m)$	$x_i(m + 0.5\ \text{kg})$	$x_i(m - 0.5\ \text{kg})$	$\bar{x}_i$	$\mid x_{i+3} - x_i \mid$	$x_{i+3} - x_i$	S	$\Delta(\Delta x)$
0.5								
2.0								
3.5								
5.0								
6.5								
8.0								

$g =$ _____ N/kg; $L_{0'} =$ _____ cm; $\Delta L = 0.1$ cm;
$d = d_F + d_R/100 =$ _____ + _____ /100 = _____ cm; $\Delta d = 0.0004$ cm;

Uncertainty in mass m (ignore the mass of the pulley and mass hanger) ± 1 g;

$$\text{Thus } E = \frac{F/A}{\Delta L/L_0} = \frac{4FL}{\pi d^2 \Delta L} = \underline{\qquad} \text{ N/m}^2.$$

$$\frac{\Delta E}{E} = \sqrt{\left(\frac{\Delta L}{L}\right)^2 + \left(\frac{2\Delta d}{d}\right)^2 + \left[\frac{\Delta(\Delta x)}{\Delta x}\right]^2} = \underline{\qquad};$$

$$\Delta E = E \cdot (\Delta E/E) = \underline{\qquad} \text{ N/m}^2;$$

$$\overline{E} \pm \Delta E = \underline{\qquad} \text{ N/m}^2$$

[Conclusion]

You are given a piece of wire hanging vertically in a sturdy frame, a two kilogram mass hanger attached to the end of the wire and a stack of kilogram masses.

The wire is connected to a bubble level and balancing micrometer, which allows you to determine small changes in the length of the wire.

You record the initial length and initial diameter of the wire.

You then place different amounts of mass on the mass hanger and for each amount you determine the cumulative change in length of the wire.

In the analysis you calculate and plot stress vs. strain.

From this graph you determine whether the stress on the wire remains in the linear region.

You find a numerical value of Young's modulus for the wire, and you determine the material that composes the wire by comparing this experimental value to a table of accepted values.

[Questions]

1. Propagate the error for the *strain* and show that it is independent of the value of ΔL. (Hint: One of the relative errors in the initial expression for $u\{e\}$ is much smaller than the other. Make the approximation that this is zero and simplify the resulting expression.)

2. One of our assumptions is that the wire diameter is constant. Show that when the actual change in area is taken into account, the fractional increase in the stress is equal to the strain (to first order). Assume that the volume of the wire is constant.

【Appendix】

Table 6 Young's moduli of various metals   GPa

Metal	E	Metal	E
Lead	15.7	titanium	120.2
magnesium	41.8	copper	129.7
aluminum-nickel alloy	64.7	platinum	166.7
cadmium	69.3	nichrome	186
silver (hard drawn)	77.5	steel (hard drawn)	192.2
antimony	78.0	steel (annealed)	200.1
gold	78.5	nickel	207.0
brass	90.2	tungsten (drawn)	355
zinc	104.7		

2. One of our assumptions is that the wire diameter is constant. Show that when the actual change in area is taken into account, the fractional increase in the stress is equal to the strain (to first order). Assume that the volume of the wire is constant.

[Appendix]

Table 6. Young's moduli of various metals

Metal	E	Metal	E (GPa)
lead	15.7	titanium	120.2
magnesium	41.5	copper	129.7
chromium-nickel alloy	64.7	platinum	156.9
cadmium	69.3	bismuth	160
silver (Hard drawn)	77.5	steel (hard drawn)	192.2
antimony	78.0	steel (annealed)	200.4
gold	78.5	nickel	207.0
brass	90.2	tungsten (drawn)	355
zinc	104.7		

3.1.6 Pendulum Measurements

[Objective]
- To illustrate an important experimental technique: isolation of variables to study the relation between a pair of variables by holding all others constant.
- To introduce graphical analysis with log-log paper as a tool for determining the relation between variables.
- To measure the acceleration due to gravity by observing the oscillation of a pendulum, to investigate the physical quantities which influence the period of a pendulum, and to determine the functional dependence of period on some of these quantities.

[Equipment]
Stopwatch; meterstick; ring stand or other support for clamp for pendulum; spring clamp; pendulum bobs of equal radius, but different masses, made of cork, wood, brass, steel, etc. ; string, one-and two-meter sticks; balance scales.

[Principle]
The simple pendulum consists of a small bob (weight) on the end of a string secured on a rigid support (the fulcrum). The bob is pulled aside a distance x and then released, allowing it to swing in an arc. The time the bob takes for one complete swing is called its period. The pendulum's period is the time it takes to swing from one side to the other and return to its starting point. This may be easily determined with a stop-clock by counting a number of complete swings, say n, and dividing the total time by n.

If we were to construct a simple pendulum by hanging a mass from a rod and then displace the mass from vertical, the pendulum would begin to oscillate about the vertical in a regular fashion. The relevant parameter that describes this oscillation is known as the period of oscillation. The period of oscillation is the time required for the pendulum to complete one cycle in its motion. This can be determined by measuring the time required for the pendulum to reoccupy a given position.

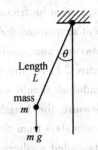

Figure 1 The pendulum

Consider a pendulum as shown in Figure 1. From a force balance we find that for small initial angles (less than 20°) the period T of the pendulum is dependent only on its length L and the acceleration due to gravity g, and independent of its mass m. That is,

$$T = 2\pi \sqrt{\frac{L}{g}}$$

It can be shown that the period of oscillation of the pendulum, T, is proportional to one over the square root of the gravitational acceleration, g. The constant of proportionality, L, depends on the physical characteristics of the pendulum such as its length and the distribution of mass about the pendulum's pivot point.

The reason that the pendulum oscillates about the vertical is that if the pendulum is displaced, the force of gravity palls down on the pendulum. The pendulum begins to move downward. When the pendulum reaches vertical it can't stop instantaneously. The pendulum continues past the vertical and upward in the opposite direction. The force of gravity slows it down until it eventually stops and begins to fall again. If there is no friction where the pendulum is attached to the ceiling and there is no wind resistance to the motion of the pendulum, this would continue forever.

Because it is the force of gravity that produces the oscillation, one might expect the period of oscillation to differ for differing values of gravity. In particular, if the force of gravity is small, there is less force pulling the pendulum downward, the pendulum moves more slowly toward vertical, and the observed period of oscillation becomes longer. Thus, by measuring the period of oscillation of a pendulum, we can estimate the gravitational force or acceleration.

[Procedure]

(1) Obtain a ring stand, pendulum clamp, and mass on the end of a string. Assemble the equipment as shown: Attach the clamp to the ring stand or other support so that it is free to hold the string for the pendulum. It is important that the pendulum suspension be friction-free so that the wood blocks or coins are used to provide a point suspension for the string. Put the blocks or coins together with the string in between, and sandwich the combination between the jaws of the clamp (see Figure 2). Tie the washers to the other end of the string to make a pendulum. The pendulum should swing freely without hitting anything.

(2) Measure the length of the pendulum. Measure the length of the pendulum in centimeters from the place where the string leaves the blocks to the middle of the suspended washers. This value should be recorded on the data table as length L for trial 1.

(3) Pull the pendulum away from the vertical at an angle of 15° (use your protractor to check the angle) and let it swing. With the stopwatch, time how long it takes to make 10 complete back-and-forth swings. Record this result on the data table. The period, T, for the pendulum is a tenth of this value.

(4) Release the clamp and change the length of the pendulum. Measure and record this new value for L on the data table. Again bring the pendulum out to 15°, and measure and record the time for 10 complete swings.

(5) Repeat step 4 three more times for a total of five trials. Each trial should use a pendulum with a different length.

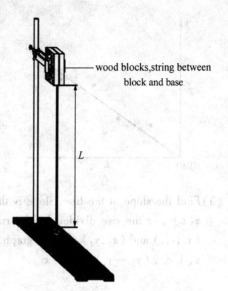

Figure 2 The setup of pendulum

(6) Complete the data table. To find the period T of one swing, divide your time values for 10 swings by 10. Then calculate and record T^2.

【Data and Analysis】

Trial	Length L/cm	Time for 10 swings/s	Period T/s	T^2/s^2
1				
2				
3				
4				
5				
6				
Avg.				

(1) Make a graph of T^2 as a function of L. Put the independent variable L on the horizontal (x) axis and the dependent variable, T^2 on the vertical (y) axis. Start the graph at the origin (0,0) and graph the five points determined by the values of T^2 and L for the six trials.

(2) Make a graph of T versus L. Connect the points in a straight line. Draw the best possible straight line through (0,0) and the six graphed points, not necessarily touching any of the points.

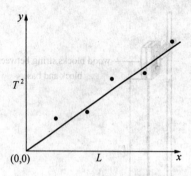 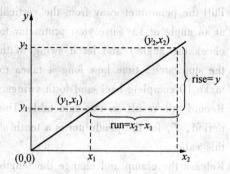

(3) Find the slope of the line. Slope is the change in y (T^2) divided by the change in $x(L)$, or the rise divided by the run. To find the slope, choose any two points, (x_1, y_1) and (x_2, y_2) on your graph. Divide the difference in y by the difference in x, i.e., $(y_2 - y_1)/(x_2 - x_1)$.

$$T = 2\pi \sqrt{\frac{L}{g}}$$

$$T^2 = 4\pi^2 \frac{L}{g}$$

(4) According to the equation for the period of a pendulum with length L, the slope that recorded above the graph may be used to find "g". The value of "g" will vary at different places on the earth.

$$\text{slope} = \frac{4\pi^2}{g}$$

(5) In any experiment, there is some degree of error. What part of the procedure do you think would be most likely to introduce some error into your results?

$$g = \frac{4\pi^2}{\text{slope}}$$

To check the effects of varies on period:

We suspect the period could depend on at least the following factors: (1) mass of the pendulum bob, (2) shape of the bob, (3) volume of the bob, (4) length of the suspension string, (5) air resistance (drag) on the bob and string, (6) size of the swing (amplitude), (7) size of the gravitational force. You have no means to change the gravitational force in our laboratory, so we must leave this variable for others to investigate, perhaps by doing the experiment on a high mountain as well as at a lower elevation. You have no practical way to investigate the effects of air resistance; but these will be small compared to experimental errors. The effect of air drag causes a noticeable slow decrease in the amplitude of swing as time goes on. The remaining factors are easily investigated.

[Procedure]

To simplify and systematize the data-taking and analysis, confine your experimental values for pendulum length to the values 15, 30, 60, 100, 200 cm or 20, 40, 80, 150, 300 cm depending on the maximum length of pendulum your laboratory situation can. Investigate at least five or six different lengths.

Use values of amplitude angle of 5, 30 and 80 degrees, measured from the vertical.

The amplitude may be expressed in several ways:
- the maximum height H attained by the bob;
- the maximum horizontal displacement X; or
- the maximum angle of displacement.

All of these are measured from the lowest point of the bob's motion. In terms of the suspension length L, these are related by $\tan \theta = X/(L - H)$. In each case, time enough swings for a time interval of at least 20 seconds, to determine the period of the pendulum's swing.

(1) Attach a string to a *rigid* support high above the floor. Suitable attachment points may be provided in the lab ceiling. The attachment point must be solid so that it does not shift position as the pendulum swings.

(2) In all cases set the pendulum into motion so it swings in a fixed vertical plane. Avoid oval paths, because they introduce other variables which are hard to control.

(3) Choose the largest practical value of string length (from the list above). Keeping this length constant, investigate the effect of bob mass on the period by successively testing each of the different bob samples provided. Use an initial amplitude of 5° in all cases.

(4) Using the same string length, and the heaviest metal bob, investigate the effect of different amplitudes. Use the amplitude angle values from the list above.

(5) Use a metal bob and an initial amplitude of 5°. Then try successively shorter string lengths. Use the length values from the list above. For the shorter lengths you will

have to time more swings to keep the accuracy of the pendulum period comparable to that for the longer lengths.

(6) If time permits, compare the periods of a plane pendulum and a spherical pendulum with the same string length and the same bob. A spherical pendulum is one in which the bob swings in a perfect circle in a horizontal plane.

[Analysis]

The data will very likely show that the strongest dependence is that of the period on the string length. Plot the data for the metal bob: period versus length, on ordinary (linear) graph paper. The plotted points will lie on a curve. The direction of its curvature leads one to suspect that an equation of the form

$$T = KL^n \qquad (1)$$

might describe the data, where K is a proportionality factor and n is an exponent less than one. We do **not** assume, as some lab manuals suggest, that $n = 1/2$, since that would be assuming something that we have not yet determined experimentally. We must find the value of n. This gives us a fine opportunity to get experience using semi-logarithmic graph paper.

(1) Take the logarithm of both sides:

$$\log T = \log K + n \log L \qquad (2)$$

(2) If this guess is correct, a plot of ($\log T$) against ($\log L$) should produce a straight line graph. Plot the T vs. L data on 2 by 1 cycle log-log graph paper. If the points do lie on a straight line, draw the best fit line with a ruler.

(3) Consider two well-separated points *on the straight line*, P_1 and P_2. Write Eq. (2) twice, once for each point.

$$\log T_1 = \log K + n \log L_1 \qquad (3)$$
$$\log T_2 = \log K + n \log L_2 \qquad (4)$$

Subtract Eq. (4) from Eq. (3).

$$\log T_2 - \log T_1 = n(\log L_2 - \log L_1)$$

So, finally,

$$n = \frac{\log T_2 - \log T_1}{\log L_2 - \log L_1} \qquad (5)$$

Use this to determine the value of n.

Also determine, from the graph, the value of K. Remember that K has units and dimensions. Express K in MKS units. From your procedure, estimate the error in each of these results.

(4) Finally, examine the data to see whether the period depends on any other variables to a degree significantly larger than the experimental errors. State clearly what can be concluded about such dependences, from the data.

(5) An elementary textbook asserts that the period of a simple pendulum is given by the equation:

$$T = 2\pi(L/g) \qquad (6)$$

where g is the acceleration due to gravity (about 9.8 m/s). How well is your data in agreement with this equation?

(6) Using your value of the slope K from analysis section (3) and Eq. (6) determine the acceleration due to gravity, g.

(7) An advanced textbook asserts that the period of a simple pendulum is dependent on amplitude angle, θ, and gives the equation for the pendulum period:

$$T = 2\pi\sqrt{\frac{L}{g}}\left[1 + \frac{1}{4}\sin^2\left(\frac{\theta}{2}\right) + \frac{9}{64}\sin^4\left(\frac{\theta}{2}\right) + \cdots\right] \qquad (7)$$

is the maximum (initial) displacement angle of the pendulum bob, measured from the vertical.

The expression within the parentheses consists of an infinite series of terms, but we have omitted those too small to concern us here. Is the data in better agreement with this equation than with the one given in part (5) above? How small must the amplitude be for the two equations to be in agreement to 1%?

【Questions】

1. Use your graph to calculate the required length of suspension which would give the pendulum a period of 12.

2. Very often the pendulum departs from motion in a single plane and begins to move in an oval path. Suppose this happened, in a pendulum swinging in an arc of 50 cm, and developed a sidewise component of motion of amplitude 10 cm. Will this affect your results? Would this alter the measured period? If so, how?

106   Conceptual Physics Experiments

$$T = 2\pi \sqrt{L/g} \qquad (6)$$

(Questions)

3.2 Thermal and Wave

3.2.1 Interference Coefficient of Thermal Expansion Measurement

The vast majority of materials have "contraction" property caused by the internal molecular thermal motion. This property should be taken into account for designing engineering structures, manufacturing machinery and equipment, and processing materials (such as welding). Otherwise, the structure stability and equipment accuracy will be compromised. The linear expansion of materials is the material thermal expansion in one-dimension. Linear expansion coefficient is an important parameter for the selection of materials. In particular, it is necessary to measure thermal expansion coefficient in new material research and development.

Here introduced a method of how to measure the metal linear expansion coefficient using interference and diffraction of light. In light interference, metal linear expansion coefficient can be measured by counting the numbers of interference pattern according to the metal temperature, while in light diffraction; it can be calculated by measuring diffraction stripe space based on the metal temperature.

【Objective】

To measure the linear expansion coefficients of metals by observing how these materials expand and contract with changes in temperature.

【Equipment】

Linear expansion apparatus consisting of steam jacket containing a metal rod.
Thermometer (reading resolution 0.1 ℃)
Temperature controller (room temperature to 80 ℃, resolution 0.1 ℃)
Dial indicator reading accuracy 0.001 mm, range 1 mm.
An automatic temperature control system which consists of contactor, voltage regulator, temperature controller, thermal resistance and heating resistor, the transmission laser beams adopts quartz optical fiber. Through the laser beams it develops the formula for the relationship between the moving number of the interference fringes and the linear thermal expansion coefficient of the object. We use the sample of special brass particle

in experiments to measure the linear thermal expansion coefficient, and greatly improve the precision.

[Principle]

Most materials expand when heated through a temperature range that does not produce a change of phase. The added heat increases the vibration kinetic energy of the atoms in the material, which in turn increases the distance between the atoms. Over small temperature ranges, the linear nature of thermal expansion leads to expansion relationships for length, area, and volume in terms of the linear expansion coefficient. In an isotropic material, the expansion occurs equally in all dimensions.

If an object of length L_1 is heated through a *small* temperature change ΔT, the change in length ΔL is proportional to the original length L_1 and to the change in temperature:

$$\Delta L = \alpha L_1 \Delta T \tag{1}$$

The proportionality constant α is called the **linear coefficient of thermal expansion**. It is clear that α is the fractional change in length per unit change in the temperature with unit in $(\text{°C})^{-1}$. Strictly speak, α varies slightly with temperature, so that the amount of expansion not only depends upon the temperature change but also upon the absolute temperature of the material. However, over the range of temperature used in this laboratory, we can assume α to be an approximately constant.

Some materials are not isotropic and have a different value for the coefficient of linear expansion dependent upon the axis along which the expansion is measured. For instance, with increasing temperature, calcite ($CaCO_3$) crystals expand along one crystal axis and contract ($\alpha < 0$) along another axis.

In this experiment, you will measure a for isotropic metal so that only along one dimension need to be measured. Also, within the limits of this experiment, α does not vary with temperature. The setup of the system is shown as below (Figure 1).

This device includes a Michelson interferometer and an oven, together with a sensor and mirrors etc. The Michelson interferometer produces interference fringes by splitting a beam of monochromatic light so that one beam strikes a fixed mirror and the other a movable mirror. When the reflected beams are brought back together, an interference pattern appears. Where, $\Delta L = L - L_1 = N\lambda/2$. Thus, $\alpha = \dfrac{\lambda}{2L_1} \dfrac{N}{\Delta T}$.

Displacement of the specimen end is measured in terms of the number of interference fringes counted.

3　CLASSICAL EXPERIMENTS

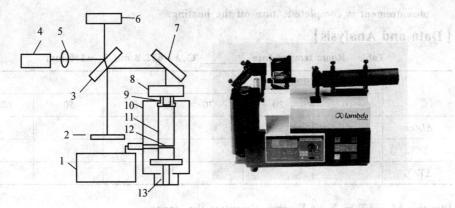

Figure 1　Schematic of system

1. Digital temperature controller　2. Viewing screen　3. Beam splitter　4. He-Ne laser　5. Beam expander
6. Fixed mirror　7. Folding mirror　8. Movable mirror　9. Quartz tube　10. Oven
11. Sample　12. Temperature sensor　13. Quartz base

From http://www.lambdasys.com/product

【Procedure】

1. Adjust the Michelson interferometer

(1) Turn on the power of laser

(2) Move beam expander (5 in Figure 1) to 90° and face outside. No laser will go through beam expander. Adjust the screw of fixed mirror and folding mirror (6,7 in Figure 1) to show a brightest overlap light point in viewing screen (2 in Figure 1).

(3) Turn beam expander (5 in Figure 1) back to its original position. Make a sharp tuning of mirrors (6, 7 in Figure 1) till a clear interference pattern appears. If there is no well-distributed beam, adjust the screws of beam splitter (2 in Figure 1).

2. Adjust the temperature control system

(1) Turn on the power and it will show current temperature.
(2) Adjust the preferred highest temperature.
(3) PAUSE and HEAT buttons can be used to stop and re-start heating during the measurement.

3. Measurement

(1) Press SELECTIVE button and set the preferred temperature to 65 ℃.
(2) Bounce SELECTIVE button and mark current temperature. Press HEAT button and record the temperature while every 10 interference cycles pass by. As soon as the

measurement is completed, turn off the heating.

【Data and Analysis】

Table   Room temperature: _____ ℃, $\lambda = 632.8$ nm, $L_0 = 150$ mm

	N					
	10	20	30	40	50	60
ΔL/cm						
T/℃						
ΔT/℃						

Plot the $\Delta L - \Delta T$ or $N - \Delta T$ curve. Calculate the slopes.

Calculate the thermal Expansion Coefficients $\alpha =$ _____ ℃$^{-1}$.

【Questions】

What are the major sources of error for the experiment? How can these errors be reduced?

3.2.2 Determining the Specific Heat Capacity of Air

【Objective】
- To measure the specific heat ratio of air by the method of adiabatic expansion.
- To learn how to use the temperature sensor and the pressure sensor.

【Equipment】
Testing Instrument is shown as below (Figure 1).

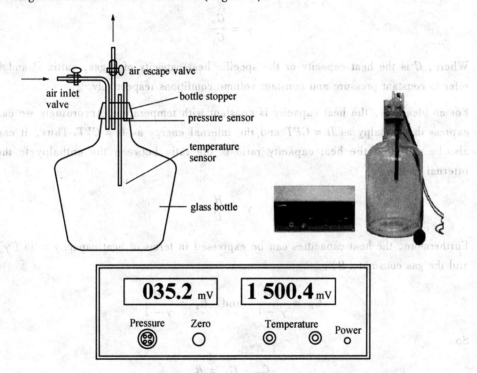

Figure 1 Testing Instrument

Sensitivity:
(1) The Pressure Sensor: 20 mV/kPa

$$P_i = P_0 + \frac{\Delta P_i}{20} (\text{kPa})$$

(2) The Temperature Sensor: 5 mV/K

$$T = \frac{U_T}{5}(K)$$

[Principle]

The heat capacity ratio or adiabatic index or ratio of specific heats, is the ratio of the heat capacity at constant pressure (C_P) to heat capacity at constant volume (C_V). It is sometimes also known as the isentropic expansion factor and is denoted by γ (gamma).

$$\gamma = \frac{C_P}{C_V}$$

Where, C is the heat capacity or the specific heat capacity of a gas, suffix P and V refer to constant pressure and constant volume conditions respectively.

For an ideal gas, the heat capacity is constant with temperature. Accordingly we can express the enthalpy as $H = CPT$ and the internal energy as $U = CVT$. Thus, it can also be said that the heat capacity ratio is the ratio between the enthalpy to the internal energy:

$$\gamma = \frac{H}{U}$$

Furthermore, the heat capacities can be expressed in terms of heat capacity ratio (γ) and the gas constant (R):

$$C_P = \frac{\gamma R}{\gamma - 1} \quad \text{and} \quad C_V = \frac{R}{\gamma - 1}$$

So,

$$C_P - C_V = R$$

The heat capacity ratio (γ) for an ideal gas can be related to the degrees of freedom (f) of a molecule by:

$$\gamma = \frac{f + 2}{f}$$

Thus we observe that for a monatomic gas, with three degrees of freedom:

$$\gamma = \frac{5}{3} = 1.67$$

while for a diatomic gas, with five degrees of freedom (at room temperature):

$$\gamma = \frac{7}{5} = 1.4$$

E. g, the terrestrial air is primarily made up of diatomic gasses (around 78% nitrogen (N_2) and around 21% oxygen (O_2)) and, at standard conditions it can be considered to be an ideal gas. A diatomic molecule has five degrees of freedom (three translational and two rotational degrees of freedom). This results in a value of

$$\gamma = \frac{7}{5} = 1.4$$

Table 1 Ratio of Specific Heats for some common gases

Gas	Ratio of Specific Heats
Carbon Dioxide	1.3
Helium	1.66
Hydrogen	1.41
Methane or Natural Gas	1.31
Nitrogen	1.4
Oxygen	1.4
Standard Air	1.4

Ideal gas law

—The state of an amount of gas is determined by its pressure, volume, and temperature according to the equation:

$$PV = nRT$$

Where P is the absolute pressure of the gas, V is the volume of the gas, n is the number of moles of gas, R is the universal gas constant, T is the absolute temperature.

The value of the ideal gas constant, R, is found to be as follows.

$$R = 8.314472 \text{ J}/(\text{mol} \cdot \text{K})$$

Table 2 Calculations

Process	Constant	Equation
Isobaric process	Pressure	V/T = constant
Isochoric process	Volume	P/T = constant
Isothermal process	Temperature	PV = constant
entropic process (Reversible adiabatic process)	Entropy	PV^γ = constant $P^{\gamma-1}/T^\gamma$ = constant $TV^{\gamma-1}$ = constant

Isotherms of an ideal gas is shown as below (Figure 2).

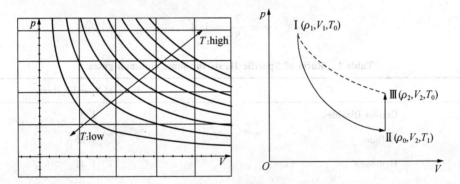

Figure 2 Isotherms

【Procedure】

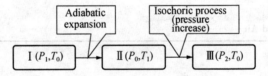

Adiabatic expansion. $\text{I}(P_1, T_0) \longrightarrow \text{II}(P_0, T_1)$

$$\frac{P^{\gamma-1}}{T^\gamma} = \text{const}$$

$$\frac{P_1^{\gamma-1}}{T_0^\gamma} = \frac{P_0^{\gamma-1}}{T_1^\gamma}$$

$$\left(\frac{P_1}{P_0}\right)^{\gamma-1} = \left(\frac{T_0}{T_1}\right)^\gamma$$

Isochoric process (pressure increase). $\text{II}(P_0, T_1) \longrightarrow \text{III}(P_2, T_0)$

3 CLASSICAL EXPERIMENTS

$$\frac{P}{T} = \text{const}$$

$$\frac{P_0}{T_1} = \frac{P_2}{T_0}$$

$$\frac{P_0}{P_2} = \frac{T_1}{T_0}$$

Through above equations, there is

$$\left(\frac{P_1}{P_0}\right)^{\gamma-1} = \left(\frac{P_2}{P_0}\right)^{\gamma}$$

$$(\gamma - 1)\ln\left(\frac{P_1}{P_0}\right) = \gamma\ln\left(\frac{P_2}{P_0}\right)$$

$$\gamma = \frac{\ln\left(\frac{P_1}{P_0}\right)}{\ln\left(\frac{P_1}{P_2}\right)}$$

【Data and Analysis】

Measurement and Data Table: $P_0 = 101.30$ kPa; $T_0 =$ _____

Trial	ΔP_1 (mV)	P_1 (kPa)	ΔP_1 (mV)	P_2 (kPa)	γ
1					
2					
3					
4					
5					
6					
Avg.					

【Result】

$$\bar{\gamma} = \frac{1}{n}\sum_{i=1}^{n}\gamma_i = \underline{\qquad} \qquad S = \sqrt{\frac{\sum_{i=1}^{n}(\gamma_i - \bar{\gamma})^2}{n-1}} = \underline{\qquad}$$

$$\gamma = \bar{\gamma} \pm \Delta\gamma = \underline{\qquad}$$

$$\gamma_0 = 1.402$$

$$E_\gamma = \frac{|\bar{\gamma} - \gamma_0|}{\gamma_0} \times 100\%$$

【Appendix】

Factors that affect specific heat capacity

For any given substance, the heat capacity of a body is directly proportional to the amount of substance it contains (measured in terms of mass or moles or volume). Doubling the amount of substance in a body doubles its heat capacity, etc.

However, when this effect has been corrected for, by dividing the heat capacity by the quantity of substance in a body, the resulting specific heat capacity is a function of the structure of the substance itself. In particular, it depends on the number of degrees of freedom that are available to the particles in the substance, each of which type of freedom allows substance particles to store energy. The translational kinetic energy of substance particles is only one of the many possible degrees of freedom, and thus the larger the number of degrees of freedom available to the particles of a substance other than translational kinetic energy, the larger will be the specific heat capacity for the substance. For example, rotational kinetic energy of gas molecules stores heat energy in a way that increases heat capacity, since this energy does not contribute to temperature.

3.2.3 Standing Waves

【Objective】

- To investigate standing waves in a string, and discover relations between measurements made on the string.
- To study the parameters which affect standing waves in various strings.
- To study the effects of string tension and density on wavelength and frequency.

【Equipment】

Long string, electrically driven vibrator, pulley, weights, weight hanger, analytic balance.

【Principle】

A wave is the propagation of a disturbance through a medium. The physical properties of that medium (e.g., density and elasticity) will dictate how the wave travels within it. A wave may be described by its basic properties of amplitude, wavelength, frequency and period T. Figure 1 displays all of these properties. The **amplitude**, A, is the height of a crest or the depth of a trough of that wave. The **wavelength**, λ, is the distance between successive crests or successive troughs. The time required for a wave to travel one wavelength is called the **period**, T. The **frequency**, f, is $1/T$, and is defined as the number cycles (or crests) that pass a given point per unit time.

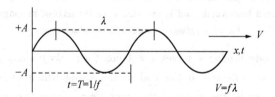

Figure 1 Properties of a standing wave

From http://www.clemson.edu/ces/phoenix/labs

Since the wave travels one wavelength in one period, the **wave velocity** is defined as λ/T. The wave velocity can then be written as

$$v = \lambda f \tag{1}$$

Relationships between these basic properties give the angular properties of the wave: $k = 2\pi/\lambda$ and $\omega = 2\pi/T = 2\pi f$, where k is the wave number and ω is the angular frequency. We can then write a one dimensional sinusoidal wave function for a wave traveling to the right as

$$y = A \cos(\omega t - kx).$$

In this experiment, we will introduce an oscillating disturbance to a length of string with the use of an electric vibrator. The vibrator shakes the string back and forth, creating a disturbance perpendicular to the string's length. This disturbance, then, propagates along the string until it hits the stationary pulley about one meter away. This wave is known as a **transverse wave** since its disturbance is perpendicular its motion.

When the wave reaches the pulley-end of the string it is reflected back toward the vibrator-end of the string. In doing so, the disturbance is not only reflected back along the string, but it is also reflected over the axis of propagation. This is shown in the Figure 2.

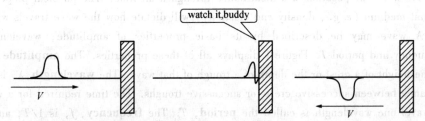

Figure 2 When a wave encounters a physical barrier such as a wall picture, it is reflected backwards and is inverted over the axis of propagation

From http://www.clemson.edu/ces/phoenix/labs

When a vibrating body produces waves $y = A \cos(\omega t - kx)$ along a tightly stretched string, the waves are reflected at the end of the string which causes two oppositely traveling waves to exist on the string at the same time. These two waves interfere with each other, creating both constructive and destructive interference in the vibrating string. If the two waves have identical amplitudes, wavelengths and velocities, a **standing wave**, or **stationary wave**, is created. That is,

$$y = A \cos(\omega t - kx) + A \cos(\omega t + kx) = 2A \cos kx \cos \omega t \qquad (2)$$

The constructive and destructive interference patterns caused by the superposition of the two waves create points of minimum displacement called **nodes** ($kx = (L + 1/2)\pi$, L is an integral number), or *nodal positions*, and points of maximum displacement called

antinodes ($kx = L\pi$). If we define the distance between two nodes (or between two antinodes) to be L, then the wavelength of the standing wave is $\lambda = 2L$. Figure 3 illustrates the case where the length of string vibrates with 5 nodes and 4 antinodes.

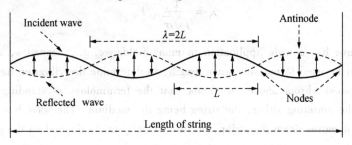

Figure 3 A standing wave is created when an incident and reflected wave have identical amplitudes, wavelengths and velocities
From http://www.clemson.edu/ces/phoenix/labs

It is possible to obtain many discrete vibration modes in a stretched string. That is, for a string to vibrate with a specific wavelength, the tension applied to the string must have a certain value. It is possible for the string to vibrate with another specific wavelength, but the tension must be adjusted until that particular mode is reached. If the tension is such that it is between vibration modes, the string will not exhibit the standing wave phenomenon and we won't see a standing wave. When the frequency of the vibrating body is the same as that of the particular vibration mode of the string, **resonance** is established. I. e. ,

$$L = (n-1)\lambda/2 \qquad (3)$$

Where n is the number of nodes.

The wave velocity of a standing wave is dependent on the medium through which the wave travels. The velocity of standing waves propagating through a taunt string, for instance, is dependent on the tension in the string, F_T, and the linear density of the string, μ. For waves of small amplitude this velocity is given by

$$v = \sqrt{\frac{F_T}{\sigma}} \qquad (4)$$

where the linear density, σ, is the mass of the string, m, divided by its length L, or

$$\sigma = (n-1)^2 \frac{F_T}{4L^2 f^2} \qquad (5)$$

Setting Equations 1, 3 and 4 equal and solving for λ, we find the relationship between wavelength and tension, namely

$$\lambda = \frac{1}{f\sqrt{\sigma}}\sqrt{F_T} \qquad (6)$$

Standing wave has a wide application in many fields, e. g., acoustics, radar, laser etc.. It can be used to measure wave length, to determine the vibration frequency of a vibrating system. From above, we know that the terminology of standing waves also applies to the vibrating string, the string being the medium. The wave has a particular velocity as it moves down the string, but this velocity might depend on the properties of the string, its mass, rigidity, tension, etc.

【Procedure】

(1) The string will be horizontal, with one end attached to the vibrator, and the other passing over a pulley to a weight hanger (Figure 4). Addition of weights to the hanger will vary the string tension. At some particular values of tension, standing waves are observed in the string. The point where the string passes over the pulley is not exactly a node, and the end attached to the vibrator is generally quite far from an antinode.

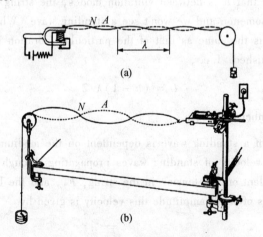

Figure 4 Vibrator driving taut string

(2) Measure a distance between n nodes which can be unambiguously located. The wavelength is twice the distance over $(n-1)$.

(3) Modify the frequency and the length of the string while keeping the weight unchanged, and form a stable standing wave. Measure a distance between n nodes.

3 CLASSICAL EXPERIMENTS

(4) Repeat step (3) four more times.
(5) Calculate the linear density of the string according to Eq. 5.

【Data and Analysis】

Tension $F_T = mg = $ _____ N

Trial No.	n	f/Hz	L/cm	$\sigma/(10^{-3}$ kg/m$)$
1				
2				
3				
4				
5				
6				

$\bar{\sigma} = $ _____ kg/m; $\Delta\sigma = \sqrt{\dfrac{\Sigma(\sigma_i - \bar{\sigma})^2}{N-1}} = $ _____ kg/m;

$\sigma = \bar{\sigma} \mp \Delta\sigma = $ _____ kg/m; $E = \dfrac{\Delta\sigma}{\bar{\sigma}} \times 100\% = $ _____

【Questions】

1. Note that tension may be varied over a considerable range without destroying the standing waves—only shifting the wavelength. But for some values of tension the string never settles down into a stable pattern. Can you discover any reason why this happens? Is it related to any of the measured quantities?

2. Are standing waves produced for any given string tension?

3. In your propagation of uncertainty, which measurement uncertainty contributed the greatest amount to the uncertainty of your final result? How could you have predicted which measurement would do so?

4. Imagine a situation where a standing wave is created on a string. Then, say the driving frequency is then doubled. Would this situation also create a standing wave? If so, how does this affect the wavelength of the standing wave?

5. A copper wire, 1.2 m long with a linear density of 0.63 g/cm, is vibrating in such a way as to produce 3 antinodes. If the wire is under a load of 350 g, what is the frequency of this mode of vibration?

6. A guitar string is plucked, creating a sound with a particular frequency. If the tension on the string is then increased by winding the string tighter, how is the string density affected? How does the changing tension and density affect the frequency of the musical note?

3.2.4 Acoustic Resonance

【Objective】

The purpose of this lab experiment is to determine the relationship between the speed, frequency and wavelength of sound waves by measuring each of these properties.

【Principle】

This laboratory experiment deals with acoustic resonance. For a more detailed treatment of waves and their basic properties, please see our standing wave experiment.

Sound waves are **longitudinal waves** which require a **medium**, such as air or water, in which to travel. Sound does not travel in a vacuum. The material properties of the medium determine the speed with which the sound wave travels through that medium. For example, the speed of sound in water is approximately 1 500 m/s and through a diamond the speed can reach 12,000 m/s! The theoretical speed of sound in dry air is given by the equation

$$v_s = 33.4 + 0.606 T_C \tag{1}$$

where T_C is the temperature in ℃, and the speed of sound is given in m/s.

In this experiment, sound waves are created with the use of speaker connected to a function generator. The **function generator** sends an oscillating current signal to the **speaker** which causes the speaker's diaphragm to vibrate. As the diaphragm moves outward, the air near the speaker is compressed, creating a small volume at relatively high pressure, which propagates away from the speaker. As the diaphragm moves inward, a low pressure area is created which also propagates away from the speaker. This is shown in the topmost cartoon in Figure 1 with the dark and light bands representing areas of high and low pressure, respectively.

The process of compressions and rarefactions continues with a **frequency** equivalent to the input signal. The higher the input frequency, the more often the compression/rarefaction cycle occurs each second. As the sound wave propagates away from the speaker, they may be detected by a microphone or a human ear. When this happens the cycle is reversed: the areas of high and low air pressure cause the ear drum and the microphone's diaphragm to vibrate, both of which create small electrical signals. In the case of the microphone, the electrical signal can be recorded by a recording device or displayed on an **oscilloscope**.

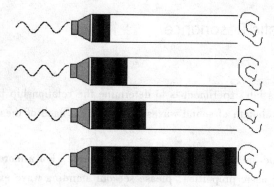

Figure 1 An alternating signal into a speaker creates sound waves which propagate away from the speaker. Here the sound waves are pumped into a resonance tube and are confined to move in one direction. The dark bands represent areas of high pressure, where the light bands represent areas of relatively low pressure.

From http://www.clemson.edu/ces/phoenix/labs

In our experiment we will confine the sound waves to move in one direction with the use of a **resonance tube apparatus**. Within the tube is a movable piston which effectively closes off one end of the tube. When a sound wave is produced by the speaker located at one end of the tube, it will propagate along the tube's volume until it hits the piston, where it is reflected back towards the speaker-end of the tube. The piston can effectively alter the length of the tube, causing the incident and reflected waves within the tube to interfere in such a way as to **maximize the amplitude** of the sound waves. When this occurs, the **standing waves** created within the tube are said to be in **resonance**. As shown in Figure 2, other **resonant modes** may be detected by slowly varying the length of the tube by moving the piston.

Note the relationship between the tube length and the sound's wavelength is given by

$$L = \frac{n\lambda}{4}, n = 1, 3, 5, 7$$

If possible, use your computer speakers or a set of head phones to listen for the resonance. If the resonance tube is closed at one end (as is the case with the tube in Figure 2), the relationship between the tube length, L, and the wavelength of the sound wave, λ, is given by the equation

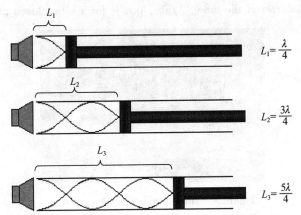

Figure 2 By moving the piston, the resonant modes may be detected

From http://www.clemson.edu/ces/phoenix/labs

$$L = \frac{n\lambda}{4}, n = 1,3,5,7,\cdots \qquad (2)$$

Notice that n is the positive **odd integers**.

The formula for resonant modes of an open tube (i.e., both ends of the tube are open) of length, L, is given by

$$L + 0.8d = \frac{n\lambda}{2}, n = 1,2,3,4,\cdots$$

A comment needs to be made here about resonance tubes open at both ends. Due to time constraints, we will not investigate standing waves in an open tube. However, students are encouraged to investigate this phenomenon on their own. It should be noted that the relationship between an open tube's length and the sound's wavelength is given by $L = \frac{n\lambda}{2}$, where n is the positive integers 1, 2, 3, 4, $\cdots$

Equation (2) does not accurately predict what actually occurs in nature. Musicians and instrument makers have known for centuries that, due to the tube's open end, an **end correction** must be made. It appears that when the sound wave exits the tube's open end, it is no longer constrained by the tube's walls. While this is true far away from the tube, the change in the character of the medium adjacent to the tube's end is significant, especially for large-diameter tubes. The corrected equation is therefore given by

$$L + 0.4d = \frac{n\lambda}{4}, n = 1,3,5,7,\cdots \qquad (3)$$

where d is the diameter of the tube. Again, this is for a tube **closed at one end**.

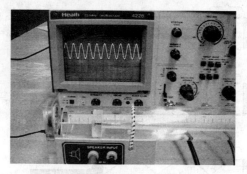

Figure 3   The resonance tube apparatus
From http://www.clemson.edu/ces/phoenix/labs

[Equipment]

The resonance tube apparatus includes a speaker, two tube supports, a piston and microphone (Figure 3). The speaker is fed a sine wave signal from the function generator, and the frequency of this sine wave should be such that at least ten data points may be taken. It is important that the signal is **not** attenuated, inverted or swept. Also, make sure that the duty cycle and dc offset is fixed.

The microphone should be secured with the thumbscrew located under the speaker. Be sure to turn the microphone on! Note that all microphones are sensitive to pressure. Therefore, when the signal from the microphone reads a maximum, this is actually a pressure **antinode** which corresponds to a **node** of the sound wave.

The function generator. The signal from the function generator is split using a T-junction. The signal is carried to the speaker via a BNC-to-banana cable and to channel 1 of the oscilloscope via a regular BNC cable.

The oscilloscope will be used to monitor both the input signal to the speaker (from the function generator) as well as the signal recorded by the microphone. To do this efficiently, you should use **both channels** of the scope.

[Procedure]

Resonance of a closed tube of varying length. In this experiment, the resonance tube speaker is being driven by a sinusoidal wave function at constant frequency. The input signal is analyzed with an FFT to ensure that it is a pure signal. The speaker-end of the tube is opened. The function generator is not shown in this picture. A moveable piston is inserted into one end of the resonance tube, thereby closing that end of the tube. As the piston is moved inside the tube, resonances can be easily heard. A microphone is inserted into one end of the tube so the resonances may also be detected by an oscilloscope.

The experiment begins with the piston pushed into the tube so that it is adjacent to the speaker. As mentioned before, the speaker is being driven at a constant frequency. The piston is then slowly moved away from the speaker, effectively increasing the tube length. When a resonance is detected the piston's position is recorded. This is repeated until the piston reaches the opposite end of the tube.

[Data and Analysis]

When resonance was achieved, the position of the piston (i. e., the tube's length) was recorded.

Times #	Length of tube, L/cm	Node # n	Frequency f/Hz	Measured tube diameter d/cm
1				
2				
3				
4				
5				
6				
Aver.				
Uncertainty				

Plot the n-L graph.

From resonance tube measurements, determine λ and d according to Eq. (3):
(1) The wavelength of the sound wave λ = _____ ;
(2) The experimental tube diameter d = _____ .

Calculate the speed of sound:
The theoretical speed of sound, v_{s0} = _____ ; The experimental speed of sound, v_s = _____
The room temperature determined according to Eq. (1), T = _____

[Questions]

1. How do you use an oscilloscope to measure voltages? The period of a AC signal?

2. Is the resonance tube an open or closed tube?

3. How do you know when resonance is achieved?

4. What steps did you take to decrease the error in this experiment?

5. Should you measure the inner or outer diameter of the tube?

6. How will you use the temperature measurement to find the speed of sound?

3.3 Optics

Light, viewed classically, is a transverse electromagnetic wave. Namely, the underlying oscillation (oscillating electric and magnetic fields) is along directions perpendicular to the direction of propagation. This is in contrast to longitudinal waves, such as sound waves, in which the oscillation is confined to the direction of propagation.

3.3.1 Measurement of the Wave Length of Light

【Objective】
- To understand the interference of light.
- To measure the wavelength of monochromatic light and different regions of the visible spectrum using a diffraction grating.
- To calculate the slit distance, grating constant.

【Equipment】

Sodium lamp, low-voltage mercury lamp, spectrometer, diffraction grating with grating constant 300 lines per mm, diffraction grating with unknown grating constant.

【Principle】

A large number of equally spaced parallel slits is called a **diffraction grating**. A diffraction grating can be thought of as an optical component that has tiny grooves cut into it. The grooves are cut so small that their measurements approach the wave length of light. A diffraction grating splits a plane wave into a number of subsidiary waves which can be brought together to form an interference pattern. See Figure 1 for details.

We know the **conditions for observable interference** are:

(1) **Coherent Sources**

Coherent sources are those which emit light waves of the same wavelength or frequency and are always in phase with each other or have a constant phase difference.

(2) **Polarization**

The wave disturbance has the same polarization.

(3) **Amplitudes**

The two sets of wave must have roughly equal amplitude.

(4) **Path Difference**

The path difference between the light waves must not be too great.

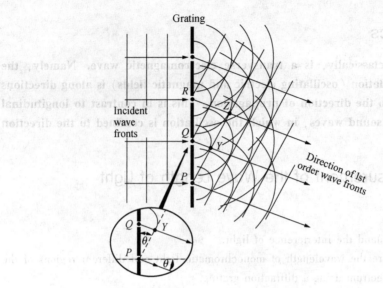

Figure 1 A diffraction grating

A diffraction grating is a series of double slits. From Figure 2, if d is the slit spacing then the path difference between the light rays X and Y is $d \sin\theta_k$. For example, if there are 600 slits per mm, then the distance, $d = 1 \times 10^{-3}$ m/ 600. The equation governing the constructive interference (principal maxima) from a diffraction grating is the same for any pair of slits.

$$d \sin \theta_k = \pm k\lambda, k = 0,1,2,3,\cdots$$

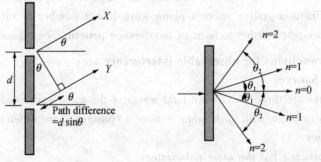

Figure 2 Path difference of two adjacent slits in a diffraction grating (left panel) and the diffraction angles of maxima (right panel)

The closer the slits, the more widely spaced are the diffracted beams. The longer the wavelength of light used, the more widely spaced are the diffracted beams. Since each monochromatic visible light leaves the grating at its own specific angle and show different color. If you measure the angle relative to k, the order of maximum, you can determine the wavelength.

Due to $\sin\theta \leq 1$, number of diffraction beams n is limited to $n \leq \dfrac{d}{\lambda}$. The highest order number is given by the value of d/λ rounded down to the nearest whole number.

We can use a diffraction grating to measure the wavelength of light. A **spectrometer** is a device to measure wavelengths of light accurately using diffraction grating to separate. The spectrometer, shown in Figure 3, consists of four components:

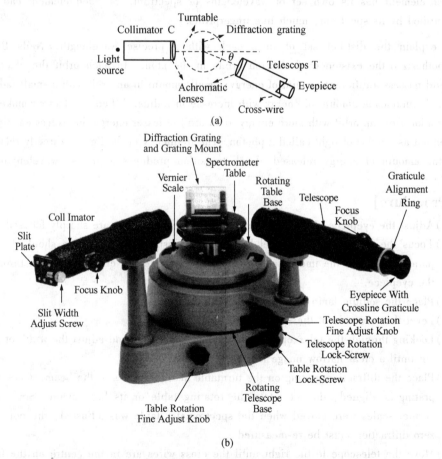

Figure 3 (a) Schematic diagram of a spectrometer, a device to measure wavelengths of light accurately using diffraction grating to separate. (b) Picture of a spectrometer

(1) A **Collimator**, for producing a collimated beam, i.e. a beam of parallel light.
(2) A **Telescope**, for viewing this light.
(3) A **Turntable**, for supporting either a *Grating* or *Prism* used to deviate the light.
(4) Two **vernier scales** to get angles between incident and output light.

Low-voltage mercury lamp: When a tube filled with rare mercury and other silent gas such as He, Ne, it is seen to glow with a purple light. When the light is sent through a diffraction grating to separate out the different colors, the purple light is found to be composed of several different colors (the color can be identified by the wavelength of the light). The separate colors appear as distinct lines which is called the **spectrum**. Each element has its own set of wavelengths or spectrum. So each element can be identified by its spectrum; much like fingerprints.

To explain the distinct set of lines each with a precise wavelength, Neils Bohr hypothesized the existence of electron orbits in the atom. In each orbit the electron would possess a different amount of energy; small amount in an orbit with a small radius and an increasing amount of energy with increase in radius. When an electron makes a transition from an orbit with more energy to an orbit of lesser energy the excess energy is released as a pulse of light called a photon. The frequency of the light is directly related to the amount of energy released. Each transition produces a precise wavelength of light.

〖 Procedure 〗

(1) Adjust the eyepiece of the telescope so that the cross-wires are sharply focused.
(2) Focus the telescope for parallel light using a distant object. There should be no parallax between the image seen in the telescope and the cross-wires seen through the eyepiece.
(3) Place the sodium lamp in front of the collimator.
(4) Level the turntable of the spectrometer if necessary.
(5) Looking through the telescope, focus the collimator lens and adjust the width of the slit until a clear narrow image is seen.
(6) Place the diffraction grating on the turntable at right angles to the beam. Once the grating is aligned, do not rotate the rotating table or its base again. Since the vernier scales were moved when the spectrometer table was adjusted, the point of zero diffraction must be re-measured.
(7) Move the telescope to the right until the cross wires are in the centre on the first bright image. Take the reading θ_R from the scale on the turntable. (To see the scale more easily shine a lamp on it and use a magnifying lens).

(8) Move the telescope back through the centre and then to the first bright image on the left. Take the reading θ_L from the scale.
(9) Calculate θ using $\theta = (\theta_R - \theta_L)/2$.
(10) Calculate the distance d between the slits using $d = 1/N$ where N is the number of lines per meter on the grating.
(11) Calculate the wavelength λ using $k\lambda = d \sin\theta_k$.
(12) Repeat this for different orders k and get an average value for the wavelength.
(13) Replace the sodium lamp by low-voltage mercury lamp in step (3) and repeat step (7) - (8) at 1st order of the spectrum. In order to cancel the error of measurement, reading the vernier scales from both left and right sides.
(14) Calculate the distance d between the slits using the known green light wavelength.
(15) Calculate the wavelength of purple, yellow lights by distance d calculated in step (14).

[Data and Analysis]

Table 1 Monochromatic light, sodium lamp

The slit distance $d =$; average $\lambda =$

k	θ_R	θ_L	$\theta = (\theta_R - \theta_L)/2$	$\lambda = d\sin\theta/k$
1				
2				
3				

Table 2 Spectrum of visible light, low-voltage mercury lamp

Spectrum diffraction angle $\theta = 1/4\,(|\theta_{-1L} - \theta_{+1L}| + |\theta_{-1R} - \theta_{+1R}|)$; $d =$

| Spectrum | θ_{-1} | | θ_{+1} | | θ | $\lambda = d\sin\theta$ |
	L	R	L	R		
Purple						
Green						
Yellow 1						
Yellow 2						

3.3.2 Single and Double-slit Interference

【Objective】
- To see the interference pattern produced by coherent light passing through single and double-slit apertures.
- To explore how these patterns depend upon the size or separation of the apertures.
- To use the interference pattern and a known slit size/separation to accurately determine the wavelength of a coherent light source.

【Equipment】

Optical bench; Laser diode w/power supply; Linear translator; Aperture bracket for light sensor; Single Slit Set; Multiple Slit Set; PASCO Xplorer (or Xplorer GLX); PASPORT Light Sensor w/cable; PASPORT Rotary Motion Sensor; Meter stick.

【Principle】

Nowhere is the wave nature of light demonstrated more clearly than in the phenomenon of interference. Many kinds of wave exhibit interference: light waves, sound waves, water waves, and so on. The underlying physics is relatively simple: when several different waves arrive at the same point in space at the same time, they pass right through each other. But at the point where the waves overlap, the total wave strength there is just the sum of the individual waves' strengths at that point. We say that these waves obey **the superposition principle**.

If we restrict our attention to just two waves at a time, and make the assumptions that these waves have the same amplitude and frequency, then we can see two limiting cases of interest to us. When the waves are **in phase**, or in other words, when the oscillations of the waves match up exactly, we can see that the waves will **constructively interfere**, and the net wave amplitude will be doubled, as shown in Figure 1(a). If, at the other extreme, the two waves are exactly 180° **out of phase**, the waves will **destructively interfere**. In this case, the net wave amplitude is zero, meaning that the waves have perfectly canceled each other out, as shown in Figure 1(b).

One way in which waves can drift out of phase with respect to each other is if they travel different distances to arrive at the same point. As a specific example, we observe that when **monochromatic**, **coherent** light passes through a narrow aperture (where narrow

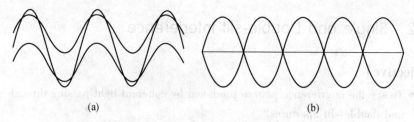

(a)   (b)

Figure 1 Superposition of equal-amplitude, equal-frequency waves. In left panel (a), the two waves are in phase, and the result is constructive interference. In right panel (b), the two waves are 180° out of phase, resulting in destructive interference

in this context implies that the width of the opening is comparable to the wavelength of the light), it spreads out into the region which we would classify as the shadow of the slit. (The shadow is the region behind the slit that light would not reach if it traveled perfectly straight lines.) The spreading of light after passing through an aperture such as this is known as **diffraction**. A top-down cartoon of this phenomenon is depicted in Figure 2 below.

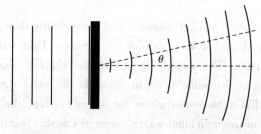

Figure 2 The spreading of a wave as it passes through an aperture whose width is comparable to the wavelength of the wave is known as diffraction. The pattern will be most intense along the forward direction (represented as a dashed line.) We will be interested in how the intensity varies as we look at various angles θ away from the center

As the light passes through each point of the opening, it spreads out in all directions, interfering, in some sense, with itself. In our minds, we imagine that each point in the opening is a new source of a wavelet of light. Due to their slightly different starting positions, each of these wavelets will travel minutely different distances to arrive at the same point in space. This will introduce a phase difference in the multitude of waves arriving at that location. Where the waves are generally all arriving out of phase with each other, the net wave amplitude will approach zero. Where the waves are generally arriving in phase, we expect the net wave amplitude to be large. With only a single slit, though, only points directly in front of the opening along the direction of propagation will generally arrive all in phase, so we expect to see a bright central spot, surrounded by fainter, alternating bands of bright and dark areas, where constructive and destructive interference occur, respectively.

Most wave sensors (*e. g.*, in humans, our sound sensors are our ears, which receive sound waves, and our eyes act as light sensors, receiving light waves) are sensitive to the intensity of the wave, rather than the amplitude. For most types of waves, **the intensity is related to the square of the wave signal**. In the present context of light waves, we relate intensity with brightness- the brighter a light source is, the higher its intensity. (For sound waves, we would say that the louder a sound is, the higher its intensity.) From the discussion above, we see that light passing through a single slit will create an alternating pattern of bright and dark areas, which we could equivalently describe as an alternating pattern of high and low intensity. We will often choose to visualize an interference pattern by showing its intensity as a function of position, for instance, of a diffraction pattern shining on a screen, as is shown in Figure 3.

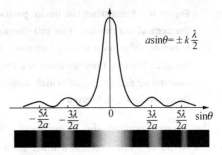

Figure 3   Intensity versus linear position of a single-slit diffraction pattern. The single-slit pattern will consist of a large, central bright spot, flanked symmetrically by alternating dark and bright areas. The central maximum will be at least ten times brighter than any of the side maxima. Where a representing the slit width, λ the wavelength of the light, and where k is an integer which labels the maximum

In general, angles are more difficult to measure than distances, and so we will often not attempt to ascertain the **angular position** of the minima directly, but will rather focus on their **linear position**, at some constant distance L away from the slit. This arrangement is typically shown in a top-down view, as is shown in Figure 4.

From that Figure 4, we can see how to relate the linear distance y of a maximum or minimum relative to the central maximum and the angle θ that the maximum or minimum makes with respect to the central maximum. Using trigonometry, we can see that they are related by $y = L \tan \theta$. Then, we can use the linear position of the intensity minima produced by a single slit of known width a in order to determine the wavelength of the light used in this experiment.

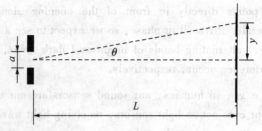

Figure 4 Measuring the linear position of the 2nd order minimum to the left of the central maximum. The interference pattern produced by the single slit is allowed to shine on a screen (or piece of paper) a uniform distance L away from the slit. We can then measure the linear location of the minima, and using trigonometry, determine at which angles, with respect to the central maximum, these minima are located

In the second portion the experiment, we will also explore the interference pattern produced by two slits, often known as **Young's double-slit experiment**, in honor of Thomas Young who performed this experiment in the early 19th century. We will illuminate two narrow slits with the same monochromatic, coherent light source as before, but we now expect to see a different pattern. First of all, more light is now going to reach our screen, and so we expect the overall pattern to be brighter (more intense). But more interestingly, we now expect to see an interference pattern due to the fact that the light from the two slits will travel different distances to arrive at the same point on the screen. If we consider two very narrow slits separated by a small distance d, the diffraction of the light from each slit will cause the light to spread out essentially uniformly over a broad central region and we would see a pattern such as the one depicted in Figure 5.

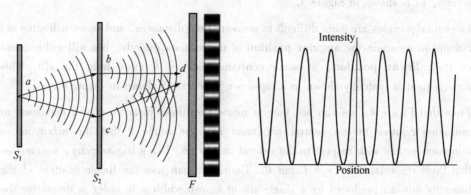

Figure 5 Idealized double-slit intensity pattern as a function of position. This is the pattern we would see if the size of the individual slits is relatively small

If the individual slits are somewhat larger, so that the diffraction patterns are not so spread out, we would expect to see a somewhat more complicated pattern, which shows both the diffraction pattern and double-slit interference pattern simultaneously, as depicted in Figure 6.

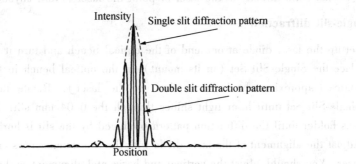

Figure 6   Actual (non-ideal) double-slit interference pattern. The maxima and minima appear where constructive and destructive interference occur, respectively, due to the path length difference between the waves propagating from each slit to the observation point (screen). This pattern is attenuated by the single-slit envelope. You can envision this pattern being created by multiplying the patterns in Figure 3 and Figure 5 together at each point along the axis

In both cases, the maxima and minima will appear where the conditions for constructive and destructive interference are satisfied, respectively. For the double-slit interference pattern, intensity maxima will be located at an angle θ relative to the central maximum, where θ will obey the relation.

$$d\sin\theta = \pm m\lambda, \qquad m = 0,1,2,3,\cdots$$

The intensity minima, on the other hand, due to the double-slit interference will occur at an angle θ relative to the central maximum given by

$$d\sin\theta = \pm \left(m + \frac{1}{2}\right)\lambda, \qquad m = 0,1,2,3,\cdots$$

Single- and double-slit interferences describe very different scenarios, and you should make every effort to keep them distinct in your mind.

【Procedure】

In this experiment, a laser diode will produce a continuous, coherent light source, which will shine through a small aperture, or slit. At the opposite end of the optical bench, a light sensor is set up that will measure the brightness (intensity) of the light shining onto that point in space. By moving the light sensor back and forth, we can see how the light intensity varies with position.

Note: Do not stare directly at the laser or point the laser toward anyone else's eyes.

1. Single-slit diffraction

(1) Set up the laser diode at one end of the optical bench and turn it on.

(2) Place the Single-Slit Set (in its mount) on the optical bench in front of the laser source (approximately 3 cm in front of the laser). Rotate the disk on the Single-Slit Set until laser light shines through the 0.04 mm slit. Rotate the black lens holder until the diffraction pattern produced by the slit is horizontal.

(3) Adjust the alignment of the laser diode with the two thumbscrews on the back of the laser. You should adjust the vertical and horizontal alignment so that the laser beam is roughly centered on the slit in the Single-Slit Set.

(4) Verify that the cable from the Light Sensor is plugged into Input 1 on your Xplorer GLX (the top-left port) and that the cable from the Rotary Motion Sensor is plugged into Input 2 (the top-right port).

(5) Prepare your GLX to take data:
   1) Press the power button on the Xplorer GLX to turn it on.
   2) Press the home button to return to the main menu.
   3) Navigate using the arrow keys, and highlight the "Data Files" icon. Select the "Data Files" item by pressing the checkmark button.
   4) Open the setup file:
      a. Arrow up to select the folders pane.
      b. Arrow right to select the "Flash" folder.
      c. Arrow down and highlight the "251_Interference" file.
      d. Press F1 to open the file.
   5) Press the home button to return to the main menu.
   6) Highlight the "Graph" icon and select it using the checkmark button. You should see a graph that is configured to display the measured light intensity plotted versus the linear position of the light sensor.

(6) On the Light Sensor, there are three buttons, each with an icon next to it representing the three sensitivity settings of the sensor. Press the button with the "candle" icon to set the sensor to its highest sensitivity.

(7) If it is not already set, turn the aperture bracket immediately in front of the Light

Sensor to position #3.

(8) Using the measuring tape on the optical bench or the meter stick, measure the distance between the single slit and the light sensor.

Record this below:

Distance between slit and sensor, L = _____.

You are now ready to take intensity measurements. You will push the light sensor on its track through the interference pattern created by the laser light shining on the slit, while the GLX records the position and intensity data. *Do your best to push the light sensor **slowly and steadily** to achieve the best results.* You may wish to practice pushing the apparatus back and forth several times to become accustomed to it.

(9) Turn off the lights in the room. Hold a sheet of white paper in front of the light sensor so that you can see the entire diffraction pattern.

Q1. Sketch below the interference pattern you see on the sheet of paper.

(10) Slide the light sensor far from the brightest spot of the diffraction pattern. Press the play button on your GLX to begin recording data, and then slowly move the Light Sensor through the pattern. Again, you should move the sensor slowly, steadily, and in only one direction. Be sure to stop the recording of data (press the play button again) when you have completed a sweep. Press F1 to auto-scale the graph.

(11) Even though you've auto-scaled the display, you'll likely need to do some fine-tuning to achieve the most useful display of the data:

1) Press F2 once to enter "Scale" mode. You can press the up arrow to increase the resolution on the vertical axis and the down arrow to decrease the resolution. Similarly, the right and left arrow keys can be used to increase or decrease the resolution on the horizontal axis, respectively.

2) Press F2 again to enter "Move" mode. Now, pressing up, down, left or right will move the axes of the graph in that direction.

3) You can get out of either Scale mode or Move mode by pressing Escape.

(12) Using the techniques described above, scale and move your graph so that the most interesting portions of the diffraction pattern are clearly displayed within the boundaries of your graph window. (Long, flat portions of the graph where the intensity was essentially constant are not very interesting.) Your graph should qualitatively match the single-slit diffraction pattern shown in the introduction (cf. Figure 3). If your graph is very rough and you cannot make out maxima and minima, re-take the data.

(13) Once you have a relatively clear graph on which you can see the minima and maxima of the single-slit diffraction pattern, we'll want to know the position of the minima so that we can determine the wavelength of the laser light. Use the data cursor and the arrow keys to locate the linear position of the various minima. **Record them in table 1**.

Table 1   Location of Intensity minima and computed wavelength in nanometers

Order	Location of Intensity Minima			Computed Wavelength/ nm
	Left of Central Max. x_L	Right of Central Max. x_R	Average: $(x_R - x_L)/2$	
$p = 1$				
$p = 2$				
$p = 3$				

Average wavelength λ = _____ nm.

Q2. Describe what happens to the diffraction pattern as the aperture becomes larger and larger.

Q3. Based on your observations, how would the diffraction pattern be affected if the slit opening were made to be much, much larger?

2. Double-slit diffraction

Step 1. Set up Multiple-Slit Set on the optical bench approximately 3 cm in front of the laser. Rotate the disk on the Multiple-Slit Set until laser light shines through the double-slit with the 0.04 mm slit width and 0.25 mm slit separation. Hold a sheet of white paper in front of the light sensor so that you can see the entire interference pattern.

Q4. Sketch below the interference pattern you see on the sheet of paper.

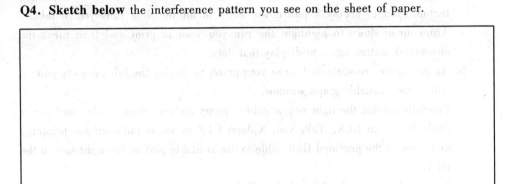

Step 2. Slide the light sensor far from the brightest spot of the interference pattern. Record data (begin by pressing the play button) while you slowly move the Light Sensor through the pattern. Again, you should move the sensor slowly, steadily, and in only one direction. Be sure to stop the recording of data (press the play button) when you have completed a sweep. Press F1 to auto-scale the graph.

Step 3. Examine your graph (by scaling and moving your graph as necessary) to see if it clearly represents the same information you sketched in Q4 above. Your graph should qualitatively match the double-slit interference pattern shown in the introduction (cf. Figure 6). If your graph is very rough and you cannot make out maxima and minima, re-take the data.

Step 4. Now rotate the Multiple-Slit disc so that the laser light shines through the slits marked "Multiple Slit (2)". This is a pair of slits, each of width $a = 0.04$ mm, separated by a distance $d = 0.125$ mm.

Step 5. Again record data (begin by pressing the play button) while you **slowly** move the Light Sensor through this interference pattern, moving the sensor steadily through the pattern, and in only direction. Be sure to stop the recording of data (press the play button) when you have completed a sweep, and press F1 to auto-scale the graph.

Step 6. You should have two clean data runs (perhaps out of a set of several runs) representing your best efforts at capturing the intensity data, for the $d = 0.25$ mm case and for the $d = 0.125$ mm case. Print and submit both of these graphs with your lab report by following these steps:

    a. To select a specific data run from the "Graph" display, press the checkmark button and arrow right to highlight the "Run #N" field. Press the checkmark button again to access a pull-down menu of all the data runs you've taken. Arrow up or down to highlight the run you wish to print and then press the checkmark button again to display that data.

    b. As necessary, re-scale and move your graph to display the full intensity pattern within the available graph window.

    c. Carefully unplug the light sensor cable, rotary motion sensor cable, and power cable from your GLX. Take your Xplorer GLX to one of the available printers, and connect the provided USB cable to the available port on the right side of the GLX.

    d. Press F4 to access the "Graphs" pull-down menu.

    e. Arrow down to highlight "Print" and select it with the checkmark button.

    f. Press F1 to begin printing.

    g. Now select the other data run requested, and repeat steps b-g to print it.

    h. Be sure to label the two printed graphs to indicate which one corresponds to the 0.25 mm case and which to the 0.125 mm case.

Step 7. When printing is complete, return the GLX to your work area, plug the light sensor into sensor port 1 (top-left), the rotary motion sensor into port 2 (top-right), and plug in the power cord.

【Data and Analysis】

Q5. Describe all the differences you've observed between the two double-slit interference patterns. (e.g. was one noticeably brighter than the other, did one have more interference peaks than the other, was one broader than the other? etc.)

Q6. How does the intensity of the brightest spots in the double-slit patterns compare to the intensity of the brightest spot in the single-slit pattern? Offer an explanation as to why this might be the case.

Q7. Based on your two graphs, what would you expect to see if you looked at an interference pattern created by two slits, each of width $a = 0.04$ mm, but separated by $d = 0.50$ mm?

Q8. Describe a method by which you could again determine the wavelength of the laser light, but now by using the double-slit interference patterns you collected. (Be specific about what you would measure, what you'd calculate, and how this gives you the wavelength of the light?)

Q6. How does the intensity of the brightest spots in the double-slit pattern compare to the intensity of the brightest spot in the single-slit pattern? Offer an explanation as to why this might be the case.

Q7. Based on your two gratings, what would you expect to see if you looked at an interference pattern created by two slits, each of width $a = 0.04$ mm, that separated by $d = 0.30$ mm?

Q8. Describe a method by which you could again determine the wavelength of the laser light, but now by using the double-slit interference patterns you collected. (Be specific about what you would measure, what you'd calculate, and how this gives you the wavelength of the light.)

3.3.3 Polarization of Light

【Objective】
- To explore the nature of polarization of light by a Polaroid polarizer.
- Verification of Malus Law.

【Equipment】

FD-OE-2 optical bench; semi-conductor laser; polavoid; quarter wavelength plate

【Principle】

The classic wave theory models light as a transverse electromagnetic wave made up of mutually perpendicular, fluctuating electric (E) and magnetic (H) fields (Figure 1). The directions of E and H and propagation follow right-hand spiral rule. The following diagram shows the electric field in the xoz plane, the magnetic field in the yoz plane and the propagation of the wave in the z direction. The figure also shows

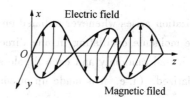

Figure 1 Electric magnetic fields of light
From http://www.pa.msu.edu/~pratts/

line tracing out the electric field and magnetic field vectors as they propagate. Traditionally, only the electric field vector is dealt with because the magnetic field component is essentially the same and optic intensity (I) is defined as modulus square of electric field A^2.

Light is said to be linearly polarized if its oscillation is confined to one direction (the direction of the oscillation of the electric field is defined as the direction of polarization). Most light sources, such as lamps and sunlight, emit unpolarized light, that is, the directions of oscillation are random. Due to the nature of transverse waves, light waves can be polarized, usually with a special filter known commercially as a **Polaroid**. A Polaroid filter blocks out light oscillating in all directions except for one, known as the direction of polarization. Additionally, unpolarized light may become polarized due to interactions with matter, via scattering and reflection.

Consider a light beam in the z direction incident on a Polaroid which has its transmission axis in the y direction. On the average, half of the incident light has its polarization axis in the y direction and half in the x direction. Thus half the intensity is transmitted, and the transmitted light is linearly polarized in the y direction.

We could use the model in Figure 2 explain polarization. The electric energy of a polarized beam acts in a specific direction that is perpendicular to the direction of propagation of the light.

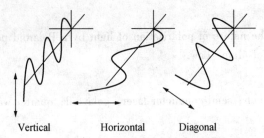

Vertical   Horizontal   Diagonal

Figure 2 The direction of polarization of the light

Polarization arises whenever radiated energy, or light, comes from an excited oscillating dipole molecule oscillating in one direction. In the natural environment these dipoles are oriented randomly and each dipole shifts its orientation over time so normal light is unpolarized. (i.e. it is made up of components of every possible polarization.)

But when light is created in a controlled environment, such as in a laser, you can selectively amplify one orientation of the polarization. Alternately, you can select one component of the polarization by
(1) passing unpolarized light through a specially designed filter.
(2) reflecting unpolarized light off an appropriately tilted surface.
(3) scattering unpolarized light in a specific direction from small particles.

Components of Polarization: Malus' Law

Like any vector, the polarization of light can always be split into a horizontal and vertical component (Figure 3).

Since the intensity is the square of the amplitude, it follows that the intensity of the horizontal and vertical components are:

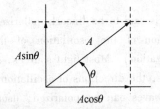

Figure 3 Components of polarization

$$I_x = A^2\cos^2\theta ; I_y = A^2\sin^2\theta$$

where angle θ is the angle between the orientation of the light and the horizontal component (or, the angle between the polarizer axis and the plane of polarization of the incoming wave).

3 CLASSICAL EXPERIMENTS

Polarization Filter, Polaroid, Polarizers

For polarized light, the electric vector oscillates in a specific direction. When light strikes a conducting material, the free electrons within the material absorb the oscillating energy and begin to oscillate themselves. The case with which the electrons can oscillate in a specific direction dictates how much light of that polarization gets absorbed.

In many metals, the molecules are amorphous, or randomly arranged, so light of all polarizations are equally absorbed. That is why metal is reflective, but not transmitted (opaque). To create a polarizer, you need to design a material that only allows oscillation of molecules in a single direction. One example is an array of thin vertical wires. If the wires are thin enough (on the order of wavelength) then electrons only oscillate in one direction and only that orientation of polarization is absorbed. The component of polarization in the orthogonal direction will transmit.

This had been difficult to do for visible light until Edwin Land invented the **Polaroid**, which is created by stretching hydrocarbon molecules on a sheet into long strands, then impregnating them with conducting iodine. In effect it acts like a set of thin wires. Light polarized in the direction of the molecules is absorbed by free flowing electrons. Light of opposite orientation is not absorbed and so is transmitted.

If a polarizer was perfect, then if one passed unpolarized light through the filter, only 50% of the light would be transmitted. For linearly polarized light, the intensity of light that is transmitted is the projection of the polarization vector onto the transmitting axis of the polarizer. This is where Malus' law becomes very useful.

First, we will establish the convention of units of light, or intensity. You can think of them as photons.

Second we describe polarizer by the orientation of their transmission axis, not the orientation of the conducting wires.

Rules for unpolarized light: 100 units of light passing through a linear polarizer, leaves 50 units of light, which will be polarized along the transmission axis of the polarizer.

Rules for polarized light: The number of units of intensity that gets through is proportional to the square of the cosine of the angle between the orientation of the incident polarization and the transmission axis of the polarizer. In other words, the transmitted intensity is the component of intensity of the original polarization that lies along the transmission axis of the polarizer.

Polarization Degree of Partially polarized light

Electric vectors of partially polarized light randomly distribute on a plane vertical to propagation direction while amplitudes of them have maximum values in some special directions and vertical to these directions show minimum values. The following figures will help us understand here mentioned definitions (Figure 4):

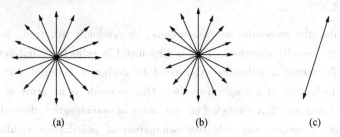

Figure 4 Different kinds of polarized light

We can understand different kinds of polarized light by this electric vector formula:

$$\vec{E} = \vec{x}E_{0x}\cos(\vec{k}\cdot\vec{z} - \omega t) + \vec{y}E_{0y}\cos(\vec{k}\cdot\vec{z} - \omega t + \phi)$$

Here ϕ is a phase shift between $\vec{x}$ and $\vec{y}$ directions. If $\phi = 0$, the electric field oscillates along a line. This is linearly polarized light. The angle of the polarization is determined by the relative magnitudes of E_{0x} and E_{0y}. If $\phi = \pm \pi/2$ and $E_{0x} = E_{0y}$, the electric field vector traces out a circle in the xoy plane. This describes circularly polarized light.

If $\phi \neq 0, \neq \pm \pi/2$ and/or $E_{0x} \neq E_{0y}$, the light is elliptically polarized. The electric field traces out an ellipse.

In this experiment, we will determine polarization degree of partially polarized light by a physical parameter P:

$$P = \frac{I_{max} - I_{min}}{I_{max} + I_{min}}$$

where I_{max} is the maximum light intensity and I_{min} represents minimum intensity value. From the formula, we can get $0 \leqslant P \leqslant 1$. If $P = 0$, it corresponds nature light while $P = 1$ means linearly polarized light.

The orientation of the emergent light is always the same as the transmission axis of the polarizing filter preceding it. So when polarizers are put in series, you can easily determine the polarization and the final emergent number of units provided that you go step by step through each polaroid, see Figure 5.

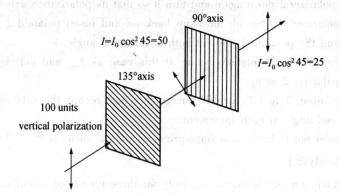

Figure 5 An example about polarized
light when polarizers are put in series

【Procedure】

1. Determination of Polarized degree of Partially Polarized Light

As followed light path figure, place a polarizer between laser and detector. Slowly rotate Polaroid P_1 and record maximum intensity I_{max} and minimum value I_{min}. Please check them for three times and calculate polarization degree P.

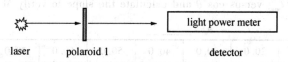

2. Verification of Malus' Law

(1) Place each polarizer on an accessory holder.

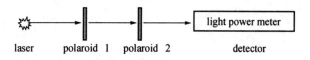

(2) Place the holder on a horizontal surface.
(3) Arrange the Light Sensor and light source so that imaginary line between the light source and the port on the Light Sensor passes through the center of the polarizers.
(4) Take polarizer 2 away and adjust the axis direction of polarizer 1 to allow the

maximum transmission of light.

(5) Then set polarizer 2 down again and turn it so that its polarization axis is parallel to the first polarizer's. (Switch the lamp back on and insert polaroid 2 between the detector and the polaroid 1 and set both at the same angle.)

(6) Record the maximum intensity value at this case as I_{max} and corresponding axis position polarizer 2 as θ_0.

(7) Rotate polaroid 2 to 90° in 10° increments and record the intensity shown in detector and angle at each increment.

Note: Be careful not to leave your fingerprints on optical surfaces.

【Data and Analysis】

(1) Record Maximum and Minimum intensity for three times and calculate polarization degree.

	1	2	3	Average
I_{max}/mW				
I_{min}/mW				

Polarization degree $P = (I_{max} - I_{min})/(I_{max} + I_{min}) = $ _____

(2) Verify Malus' Law.

Plot a graph of I/I_{max} versus $\cos^2\theta$ and calculate the slope to verify Malus' Law.

$I_{max} = $ _____ $\theta_0 = $ _____

Angle/(°)	10.0	20.0	30.0	40.0	50.0	60.0	70.0	80.0	90.0
$\cos^2\theta$	0.970	0.883	0.750	0.587	0.413	0.250	0.117	0.030	0.000
I/mW									
I/I_{max}									

3.3.4 Sugar Identification Using Polarimetry

【Introduction】

Polarimetry is the measurement and interpretation of the polarization of transverse waves, most notably electromagnetic waves, such as radio or light waves. Typically polarimetry is done on electromagnetic waves that have traveled through or have been reflected, refracted, or diffracted by some material in order to characterize that object.

Optically active samples, such as solutions of chiral molecules, often exhibit circular birefringence. Circular birefringence causes rotation of the polarization of plane polarized light as it passes through the sample. Most organic macromolecules possess chirality (not identical to its mirror image), usually called left-rotated and right-rotated molecules. In fact, all the amino acids of human bodies are left-rotated and only by artificial synthesis methods can get corresponding right-rotated amino acids. Yet, the synthesized materials may bring fatal dangers to our bodies. A worst case is the one that the non-desired enantiomer (nonsuperimpossible isomers that the overall three-dimensional configuration a chiral molecule cannot be superimposed on its mirror image) causes serious toxicity. In fact, all chiral materials may result in optical rotation and non-chiral ones have no this characteristic.

Polarimetry is a sensitive, nondestructive technique for measuring the optical activity exhibited by inorganic and organic compounds. A compound is considered to be optically active if linearly polarized light is rotated when passing through it. The amount of optical rotation is determined by the molecular structure and concentration of chiral molecules in the substance. The polarimetric method is a simple and accurate means for determination and investigation of structure in macro, semi-micro and micro analysis of expensive and non-duplicable samples. Polarimetry is employed in quality control, process control and research in the pharmaceutical, chemical, essential oil, flavor and food industries. It is so well established that the United States Pharmacopoeia and the Food and Drug Administration include polarimetric specifications for numerous substances.

Here, the experiment will demonstrate us how to determine chirality as well as concentration of sugar solution using polarimeter.

【Objective】
- Understand optical rotation phenomenon and try to detect chirality as well as concentration of sugar solution using half-shade polarimeter.

- Use optical rotation as a method of determining the identity of unknown sugars.

【Equipment】

WXG-4 polarimeter, tubes for containing sugar solution, natrium lamp.

【Principle】

1. Optical rotation

Chiral molecules have an asymmetrical center which respond to light as a lens and rotate the light. The ability to rotate light is termed as optical activity. Enantiomeric compounds rotate light by exactly the same amount but in the opposite direction. The degree to which a substance rotates light may be used to determine:

(1) the identity of the substance, the enantiomeric purity of the substance or
(2) the concentration of a known substance in a solution.

In order to observe rotation, the light which is passed through the solution must be plane polarized. From Figure 1, ordinary light has waves which are oriented in all directions, plane polarized light is made up of waves which are oriented parallel to a defined plane.

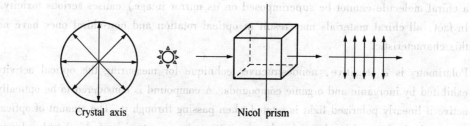

Figure 1 **Ordinary light and plane polarized light**

When a beam of plane polarized light passes through a solution of optically active material the light will rotate, see figure 2.

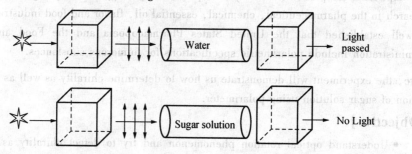

Figure 2 **Plane polarized light passes through a solution of optically active material**

3 CLASSICAL EXPERIMENTS

Each pure chiral material has a set specific rotation θ which is a fixed physical characteristic for that material. The specific rotation angle describes how far a material will rotate light. $\theta = 60°$ means that pure material will rotate light in the clockwise manner by 60 degrees. It is also concluded that the material is right-rotated. The enantiomer (see figure 3 for details) will rotate the plane of polarized light by exactly the same amount but in the opposite direction. $\theta = -60°$ would result in a rotation in the counterclockwise direction by 60 degrees. It is noted that when equal parts of two enantiomers are mixed, there will have no net rotation because the equal but opposite rotations cancel each other.

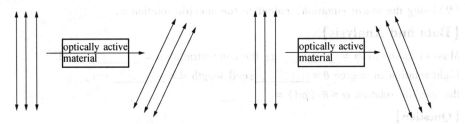

Figure 3 Right-rotated optic phenomenon and left-rotated optic phenomenon

2. The concentration of a known substance in a solution

Experiments show the rotation angle θ is related to the property and thickness d of chiral material as well as the wavelength of incident light:

$$\theta = \alpha \rho d$$

Here, α is specific rotary power (specific rotation of the compound), ρ is the concentration of solution (concentration in grams per milliliter (g/mL)), θ is observed rotation of light in degrees, d is cell length in decimeters (dm).

In this experiment a sugar solution of known concentration ρ, but unknown identity will be prepared (note the units of concentration used). The observed rotation θ will be obtained by using a polarimeter. This data will be used to calculate the specific rotation α and the identity of the sugar will be hypothesized.

[Procedure]

(1) Weigh out approximately 0.5 g of unknown sugar. Record all the digits of this mass in your notebook.
(2) In 50 mL beaker, dissolve your material in approximately 10 mL of deionized water. Swirl the contents until all the solid has dissolved.
(3) Carefully transfer this solution to a 25 mL volumetric flask.

(4) Rinse beaker with approximately 2 mL of deionized water. Transfer this solution to volumetric flask. Repeat.
(5) Carefully drop wise add deionized water to volumetric flask until bottom of meniscus is exactly on line. This is your solution.
(6) Calculate the concentration ρ in g/mL.
(7) Obtain the θ by analyzing your solution in the polarimeter using the instructions mounted adjacent to machine and reviewed by your instructor.
(8) After an acceptable measurement is obtained, empty volumetric flask, rinse volumetric with deionized water, and return volumetric flask to supply bench.
(9) Using the above equation, calculate the specific rotation α.

【Data and Analysis】

Mass of unknown m = _____ g; the concentration ρ = _____ g/mL
Light rotation in degree θ = _____ ; cell length d = _____ dm
the specific rotation $\alpha = \theta/(\rho d)$ = _____

【Question】

Temperature can affect the rotation of light, which should be accounted for in the calculations. Can you design an experiment about polarimeter to show the temperature influence in measuring chirality and concentration of sugar solution?

3.4 Electricity

3.4.1 Measurement of Electromotive Force with Compensation Method

【Objective】
- Master the principle and method of measuring electromotive force with compensation method.
- Understand the principle of structure and application of potentiometer.

【Equipment】
String potentiometer, steady power supply, galvanometer, *emf* sources; *emf* standard; dual-switch.

【Principle】
Every current source redistributes electric charges in an electric circuit. A battery is called either a source of electromotive force or, more commonly, a source of *emf*. The *emf* of a battery is the maximum possible voltage that the battery can provide between its terminals. When an electric potential difference exists between two points, the source moves charges "uphill" from the lower potential to the higher. Consider the circuit consisting of a battery and a resistor. The positive terminal of the battery is at a higher potential than the negative terminal. Because a real battery is made of matter, there is resistance to the flow of charge within the battery. This resistance is called internal resistance r. For an idealized battery with zero internal resistance, the potential difference across the battery (called its terminal voltage) equals its *emf*. However, for a real battery, the terminal voltage is not equal to the *emf* for a battery in a circuit in which there is a current.

Thus the potential difference across the battery (terminal voltage) is always smaller than its electromotive force of the battery. The relationship of electromotive force (E), terminal voltage ($U_A - U_B$, potential difference) between electrodes and internal resistance (r_i) is as following,

$$E = (U_A - U_B) + Ir_i$$

The formula above indicates that if current I becomes smaller and smaller, the value of terminal voltage $U_A - U_B$ becomes closer and closer to electromotive force. If current I is zero, the electromotive force E will be equal to the terminal voltage, that is, $E = U_A - U_B$. In this formula, r_i is the internal resistance of battery.

This is the key point of the principle of this experiment to measure the electromotive force of batteries. To make it clear, by avoiding current passing through the battery to minimize extremely the potential descending inside battery, we can get the electromotive force value by measuring the potential difference between electrodes.

How to make it possible that there is no current passing through battery? The most popular method is compensation method. The diagram of principle of compensation methods is shown in figure 1. E_0 is adjustable power supply, E_x is the power supply needed to be measured. Place the two power supply as this way: positive electrode vs. positive electrode, negative electrode facing negative electrode. Adjust the power supply E_0 to get "zero" current on the galvanometer, thus $E_x = E_0$. The circuit is under compensation state now. If E_0 is known, E_x will be the same value. This method is called compensation method to measure the electromotive force by application of compensation principle.

Figure 2 is the diagram of principle of measurement of unknown electromotive force. Power supply E and accurate resistor R_{ab} is connected in series to form a closed circuit. When there is constant current I_0 passing through R_{ab}, the potential difference between c, d can be varied by adjusting the positions of c, d. The value of V_{cd} is proportional to the resistance between c and d. In practice, resistance is converted into potential difference based on I_0 is constant. At first, throw the switch to E_s side, when the current flowing through galvanometer is zero, then V_{cd} equals to E_s. $V_{cd} = E_s = IR_{cd}$. Then throw the switch to the E_x side, adjust the position of c and d to new positions c' and d', till the current through galvanometer is zero, in this case $V_{c'd'} = E_x = IR_{c'd'}$. **Note to make sure that positive electrodes are connected to positive electrodes, negative to negative electrodes among E_s, E_x, and E in the circuit.**

Resistance of a wire with constant cross-section is given by formula $R = \rho \dfrac{L}{S}$, where ρ is resistivity of material, L is length of the wire and S is its cross-section area. Because current through the resistor ab is constant, the electromotive force E_s is proportional to the R_{cd}. The same case is for E_x and $R_{c'd'}$. Measure the value of L_s, L_x when the current in corresponding sub-circuit is zero. Then, $E_s = I\rho \dfrac{L_s}{S}$ $E_x = I\rho \dfrac{L_x}{S}$

Division of the two formulas above, the following is got,

$$E_x = E_s \frac{L_x}{L_s}$$

E_s is the electromotive force of standard battery with a known value, so the E_x value of the battery checked can be deduced by formula above.

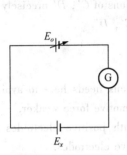
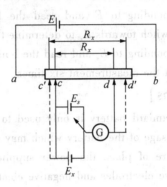

Figure 1 The diagram of principle of compensation methods

Figure 2 The diagram of principle of measurement of unknown electromotive force

The diagram of the structure of apparatus in this experiment is shown in Figure 3. The resistor wire AB is folded into 11 segments with the length 1 m of each segment and put onto a plate evenly. Touch point C is movable by inserting into any hole of 1, 2, 3, 4, ⋯ to find a proper position. There is also a meter stick under the bottom segment of the wire. The movable touch point D on the resistor wire can move to the left or to the right to find the balance position. The dual switch can be flipped up or down to connect to E_s or E_x.

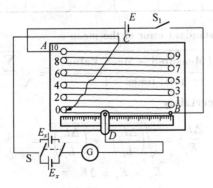

Figure 3 Structure of apparatus

【Procedure】

(1) Connect the circuit correctly according to figure 3 and ask the supervisor to check it before operating.

(2) Throw the dual-switch towards E_s, practice many times to grasp the method to find balance position. It must be done this way, first find C', then D'.

(3) Operate the measurement now. Find the balance position of C, D accurately where corresponding to E_s and read the length L_s between C, D. Then throw the dual-switch towards E_x to determine the balance positions of C', D' precisely where corresponding to E_x and read the length L_x between C', D'.

(4) Repeat the measurement six times.

【Cautions】

(1) The standard battery is only used to compare. Cautions needs here to avoid long term usage of the battery which may cause its electromotive force weaker.

(2) Be sure of place the power supplies E, E_s, E_x with positive electrodes facing positive electrodes and negative electrodes vs. negative electrodes.

(3) Do not slide the touch point D along the wire to find the balance point. Tapping mode should be applied to keep the wire from degeneration.

【Data and Analysis】

1. Data Recording

$E_s = $ _____ V $E = $ _____ V $\Delta_B = $ _____ m $\Delta L = (\Delta_A^2 + \Delta_B^2)^{1/2}$

	1	2	3	4	5	6	$\bar{L}$	Δ_A	ΔL
L_s/m									
L_x/m									

Where Δ_A equals to the standard error of the mean

$$S = \sqrt{\frac{\sum(\text{Measured} - \text{Mean Value})^2}{\text{total times} - 1}} = \sqrt{\frac{\sum(x_i - \bar{x})^2}{n-1}}$$

2. Calculate E_x and Uncertainty

$\bar{E}_x = \bar{E}_s \dfrac{\bar{L}_x}{\bar{L}_s} = $ $\dfrac{\Delta E_x}{E_x} = \sqrt{\left(\dfrac{\Delta L_s}{L_s}\right)^2 + \left(\dfrac{\Delta L_x}{L_x}\right)^2} = $ $\Delta E_x = \dfrac{\Delta E_x}{E_x} \bar{E}_x = $

Results $E_x = \bar{E}_x \pm \Delta E_x = $

【Questions】

1. What is compensation method? What is the advantage of it?

2. What would be the possible reasons in the following situation? The circuit was connected based on figure 3. After the dual-switch was flipped towards E_x or E_s, there was always one-side turn of the pointer of galvanometer no matter what adjustment applied to movable touch point C, D.

3. What needs to pay attention to when you operate a standard battery?

[Questions]

1. What is compensation method? What is the advantage of it?

2. What would be the possible reasons in the following situation? The circuit was connected based on figure 3. After the dual-switch was flipped towards E or E_x, there was always one-side turn of the pointer of galvanometer no matter what adjustment applied to movable touch point C, D.

3. What needs to pay attention to when you operate a standard battery?

3.4.2 Direct Current, Alternating Current Measurement and Ohm's Law

【Objective】
- To understand some usage of basic electrical instruments, such as ammeter, voltmeter, multimeter, etc.
- To measured DC and AC voltage and current with multimeter.
- To understand how to correctly use the meter in a circuit to obtain the correct reading.
- To verify the Ohms' law.

【Equipment】
Multimeter, ammeter, voltmeter, experimental circuitry board.

【Principle】
Electrical meters are devices that measure important physical properties of electrical circuits. They can measure electrical component properties such as resistance of resistors, inductance of inductors, and capacitance of capacitors. They can also measure active properties of a circuit during its operation such as voltage and current. Our multimeter can function as a voltmeter (AC and DC), an ammeter (AC and DC), an ohm meter and a diode checker.

Current is the flow of charge through a device. In our case, this means the flow of electrons in a circuit. The unit of current is the Ampere. An ammeter measures the flow of charge (current) through a wire or electrical device. Two important facts about ammeters are:
(1) Ammeters have very low resistance to the current so that, when placed in a circuit, the current is not impeded.
(2) Ammeters have to be in series with the circuit whose current is to be measured since they measure the flow of charge.

Voltage differences are generated by devices such as batteries (chemical energy), and electrical generators (magnetic), and photovoltaic cells (solar). Voltage is a measure of the work per unit charge required to move a unit charge between two points. The unit of voltage is the volt. A voltmeter measures the difference in electrical potential, called the "voltage drop" (or usually the "voltage across" a device), between two points in a circuit. Two important points about voltmeters are:

(1) A voltmeter has very high resistance to the flow of electricity so that, when it straddles two points in a circuit, a new path for the current is not created.

(2) Voltmeters must go in parallel with the component being measured because of this.

Ohm's law for direct current circuit states that the voltage difference V between two points in a conducting medium is proportional to the current I between those points. That is,

$$V = IR$$

where the resistance R is a constant of proportionality. Ohm's law states, in other words, that the resistance between the points is constant.

1. The Structure of Multimeter

Multimeters may use analog or digital circuits—analog multimeters (AMM) and digital multimeters (often abbreviated DMM or DVOM). Analog instruments are usually based on a microammeter whose pointer moves over a scale calibration for all the different measurements that can be made; digital instruments usually display digits, but may display a bar of a length proportional to the quantity measured.

The principles of direct current ammeter (DCA), direct current voltmeter (DCV) and resistor meter are shown in figure 1. Both the ammeter and voltmeter are basically galvanometers. The electrical resistance of the instrument is made to be very low for the ammeter and very high for the voltmeter. In DCA measuring circuit, the internal resistor is about kΩ, sensitive degree is 10 μA, and $R_1 \sim R_5$ are aimed to reduce the current of DCA. The values of these resistors are often smaller than several ohms. In DCV measuring circuit, the function of $R_6 \sim R_{10}$ is to increase the total resistance (Figure 1).

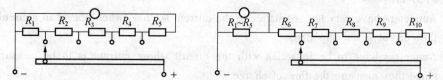

Figure 1　The measure circuit of DCA and DCV

2. Root Mean Square (RMS) Values of alternating current and voltage

AC voltage and currents cannot be measured with a DC-meter because the pointer cannot change direction quickly enough. They can be measured either with an AC meter. The magnitude of an AC voltage may be characterized by the amplitude V_{max}, but a more common practice is to use the "**root-mean-square**" voltage ("**effective value**"), which is indicated by V_{rms}. The "root-mean-square" voltage depends on the type of the AC signal. For an AC sinusoidal signal (see Figure 2): $V_{rms} = V_{max}/\sqrt{2}$.

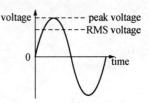

Figure 2 **The peak voltage and RMS voltage**

In general, any measured value of current or voltage in any AC application is, in fact, an effective value unless otherwise specified. AC voltmeters and ammeters show the **RMS value** of the voltage or current.

3. To check fault in the circuit multimeter

(1) Measuring voltage: the potential difference between two points has certain value in normal state. When there is a fault in circuit, the potential difference between two points will change. So according to the distribution of potential difference, the points in fault will be determined.
(2) Measuring resistance: make sure to turn off the power, and then measuring the resistance of branches with multimeter. According to the distribution of resistance value, the fault points will be determined.

【Procedure】

1. The measuring principle

(1) During measuring, the black probe of multimeter should be plugged into "COM" jack, meanwhile the red probe should be plugged into "10 A" or "mA" jack (depending on the measuring current). When resistance, voltage or capacitance is measured respectively, the red probe should be plugged into "V/Ω" jack.
(2) When the value of to be measured quantity is unknown, the switch should be placed at the maximum measured scale.
(3) Digital instruments usually display digit, the quantity measured is related to the choosing unit and measured scale. For example, if the scale of 2M is used to measure resistance, the measuring value is 0.001, so the resistance is 0.001 MΩ.
(4) Before changing measuring scale, the probe should be removed from the circuit. During measuring, do not contact the probe tip by hand.
(5) Pay attention to multimeter should be connected in a parallel circuit when voltage is measured; meanwhile multimeter should be connected in a series circuit when current is measured. A digital instrument is shown in Figure 3.

Figure 3 A digital multimeter instrument

2. The methods of measuring electric quantities

(1) Measuring direct current resistance.

The black probe of multimeter should be plugged into "COM" jack, meanwhile the red probe should be plugged into "V/Ω" jack. And turn the dial to the Ω setting. In Figure 4, resistance of resistor 1 and 2, resistance of bulb, the direct current resistance of primary and secondary coil will be measured. Caution: Switch off A and B, the resistance between power supply two terminals equals the direct current resistance of primary coil. The resistance between points B and D equals the direct current resistance of the secondary coil.

(2) Measurement of the property of diode and capacitance.

The black probe of multimeter should be plugged into "COM" jack, meanwhile the red probe should be plugged into "V/Ω" jack. Turn the dial to the diode properly measured setting. And make the other end of red probe connected to the positive pole of diode, meanwhile the other end of black probe connected to the negative pole of the diode. Now the multimeter displays the value of positive potential difference (0.5 V). Exchange the places of two probes, and measure it again, the multimeter displays 1, which means in good function.

(3) Measurement of the electromotive force of battery.

The black probe of multimeter should be plugged into "COM" jack, meanwhile the red probe should be plugged into "V/Ω" jack. To use the multimeter as a voltmeter in this lab, we want to turn the dial to Voltage setting. Then measure the electromotive force of two different batteries.

3. Measurement of the direct current circuit

The direct current power supply is connected into the circuit as shown in Figure 4. In the case of disconnecting A and C, terminal voltage of power supply is measured, which equals the electromotive force of this power supply. And connect A and C, repeat the above measurement.

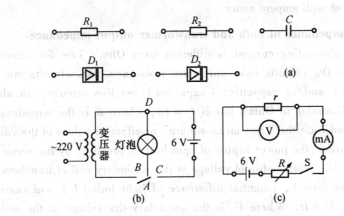

Figure 4 The measure circuit of resistors

Turn the multimeter into ampere meter. The red probe should be plugged into "mA" jack. Turn the dial to the direct current 200 mA setting. Disconnect A and C, measure the circuit current with ampere meter.

4. Verify the ohm's law in direct current circuits

(1) Add power supply (6 V), multimeter and voltmeter into circuit, and make the voltage values as 1 V, 2 V, 3 V, 4 V, 5 V by adjusting the resistor. The different current values will be recorded.

(2) Draw the V-I curve (it should be a line). **Note:** make current I as horizontal axis, voltage as vertical axis.

(3) Ultimately calculate the resistance through its slope.

5. Measurement of AC circuit

Measure the AC voltage:

Plug one end of the black probe of multimeter into "COM" jack, meanwhile plug the red probe of multimeter into "V/Ω" jack, and then turn the dial to the "alternating voltage 20 V" setting. The alternating current power supply (220 V), that is, the primary coil of transformer, is connected into the circuit as shown in Figure 4. In the case of disconnecting A and B, measure the voltage (U_0) across the secondary the voltage of the coil without bulb. Next connect A and B, repeat this measurement procedure, the alternating potential difference across bulb (U) will be measured.

Measure the AC current:
Turn the multimeter into ampere meter. The red probe should be plugged into "mA" jack. Turn the dial to the direct current 200 mA setting. Disconnect A and C, measure the circuit current with ampere meter.

Calculate the impedance of bulb and transformer output impedance:
Ohms Law for alternating current is different from Ohm's Law for direct current circuits because DC circuits have only pure resistance. AC circuits may contain Inductive (coil) and/or capacitive (capacitor) reactive circuits. In alternating current, the AC analog to Ohm's law is $U = IZ$, where Z is the impedance of the circuit and U and I are the "root-mean-square" or effective values of the voltage and current. In figure 4, the power supply of light bulb is provided by the secondary coil of transformer, The relationship of voltage of the secondary coil of transformer (U_0, AC electromotive force), potential difference of light bulb (U) and current is as following, $U_0 = U + Iz$. Where U_0 is the secondary the voltage of the coil without load bulb, U is the alternating potential difference across bulb. z is the impedance of the transformer output. U and I are the "root-mean-square" or effective values of the voltage and current.

【Data and Analysis】

1. The Methods of Measuring Electric Quantities

(1) Measuring direct current resistance

resistance	R_1	R_2	Bulb	Transformer (1st coil)	Transformer (2nd coil)
measurement					

(2) Diode and capacitor

Diode 1 Voltage/V		Diode 2 Voltage/V		capacitor
Positive terminal	Negative terminal	Positive terminal	Negative terminal	Capacitance/μF

(3) Electromotive force of battery

Battery	E_1	E_2
Electromotive force/V		

2. Directive Current Resistance and Ohm's Law

(1) Directive current circuit

E/V	U/V	I/mA	$R = U/I/\Omega$	$r = E/I - R/\Omega$

(2) Verify Ohm's law

U/V	1.00	2.00	3.00	4.00	5.00
I/mA					

Plot U-I curve and find the slope to verify Ohm's law.

3. Alternating current resistance and Ohm's law

Recording the voltage U_0 (the secondary coil of transformer), U (the alternating voltage of bulb), alternating current I, the impedance Z of the light bulb. Then, calculate the impedance of the transformer output z according to $z = U_0/I - Z$.

U_0/V	U/V	I/mA	$Z = U/I/\Omega$	$z = U_0/I - Z/\Omega$

【Questions】

1. In which condition does multimeter have the maximum (or minimum) internal resistance?

2. When the current flows through the circuit, can a resistance be measured by a multimeter? Why or why not?

3. How to check the circuit by using a multimeter?

4. What are the differences between direct current and alternative current?

5. What is the physical meaning of RMS value? For example, the value of voltage or current?

3.4.3 The Features and the Use of the Digital Oscilloscope

【Objective】
- To understand the fundamental mechanism, and usage of digital oscilloscope.
- To observe various waveforms of voltage and to measure their voltage and frequencies.
- To observe Lissajous'curve and beat.

【Equipment】
A Rigol DS1102E Digital oscilloscope; a low frequency signal generator; signal source.

【Principle】

1. Introduction

Waveforms

A waveform is a representation of how alternating current (AC) varies with time. Perhaps the most familiar AC waveform is the sine wave, which derives its name from the fact that the current or voltage varies with the sine of the elapsed time. Common uses for sine waves are in wireless technology. An ideal, unmodulated wireless signal has a sine waveform, with a frequency usually measured in megahertz (MHz) or gigahertz (GHz). Household utility current has a sine waveform with a frequency of 50 Hz in our country; most countries including the United States, it is 60 Hz. A waveform consists of various properties, two of which are amplitude (A) and period (T). Other common AC waveforms are the square wave, the ramp up and the ramp down waves (both are also known as a saw-tooth wave), and the triangular wave.

Oscilloscope

A device that displays the waveform created by an electrical device, such as a frequency generator or a specific circuit, is known as an oscilloscope. The "trace" is generated on the screen in a specific manner so that the amplitude and period can be measured. Analog oscilloscopes used a cathode ray system that plotted the trace on a phosphorus screen. To do this, it used magnetic plates to control the beam, much like non-plasma computer monitors and TVs. The beam is "aimed" at a particular spot on the phosphor screen by two pair of electrostatic deflection plates, one set to deflect the beam up or down and the other to deflect the beam right or left. The deflection plates' potentials are controlled by the output of two independent electronic amplifiers. We'll refer to these as the "vertical" and the "horizontal" amplifiers.

Digital scopes, on the other hand, sample the signal at a very fast rate, then plot the results on a graph. The digital storage oscilloscope, or DSO for short, is now the preferred type for most industrial applications, although simple analog cathode ray oscilloscopes (CROs) are still used. We will use digital scope in this experiment.

The amplitude of the waveform can be read from the scope. The volts/div setting tells us how many volts each vertical division on the graph means. So, if we have a wave that is 4 divisions high and the volts/div is set at 2 V/div, the waveform's amplitude is 8 V. Likewise, we can read the period of the waveform from the time/div setting on the horizontal axis. If the setting is 2 ms/div (two milliseconds per division), and the number of divisions from trough to trough is 3, the period of the wave is 6 ms. To convert this to frequency in Hz, take the inverse: $T = 1/f$.

Modern oscilloscopes often have two independent amplifiers which may be used simultaneously. Each is called a "channel" and such an oscilloscope is a "dual channel" oscilloscope. Dual channel oscilloscopes can display two different signals simultaneously on the screen, and may have the capability to display their sum or their difference.

The Rigol DS1102E Model oscilloscope that we will use is economical digital oscilloscope with high-performance, which is designed with dual channels and 1 external trigger channel. It provides 20 types of wave parameters for automatically measuring, which contains 10 Voltage and 10 Time parameters (Figure 1).

In front panel, the main region can be divided into USB host, LCD, Mode/Functions Buttons, Direction Key and knob, CH1 output, CH2 output/Counter Input, Power Button, Local View, Menu Buttons, Wave Selection Buttons, Channel Switch, Keyboard etc..

2. The Basic Operating Method of Digital Oscilloscopes

(1) To Compensate Probes

Perform this adjustment to match the characteristics of the probe and the channel input. This should be performed whenever attaching a probe to any input channel for the first time. From CH1 menu, set the Probe attenuation to 10 × (press CH1→Probe→10 ×). Set the switch to 10X on the probe and connect it to CH1 of the oscilloscope. When using the probe hook-tip, inserting the tip onto the probe firmly to ensure a proper connection. Attach the probe tip to the Probe compensator connector and the reference

3 CLASSICAL EXPERIMENTS

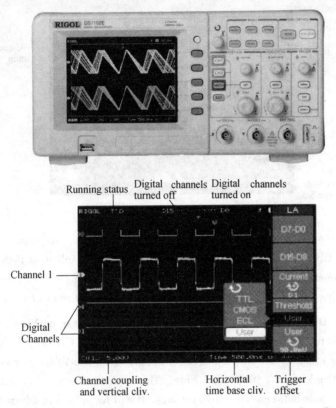

Figure 1 Rigol DS1102E Digital Oscilloscopes and its User Interface
From http://www.rigol.com/down load

lead to the ground pin, Select CH1, and then press AUTO.

(2) **To Display a Signal Automatically**

The oscilloscope has an automated feature to display the input signal best-fit. The input signal should be 50 Hz or higher and a duty cycle is greater than 1%. Pressing the AUTO button, the oscilloscope automatically sets up VERTICAL, HORIZONTAL and TRIGGER controls to display the input signal. Adjust the controls manually to get the best results if necessary.

(3) **To Set up the Vertical System**

1) **Settings of the Channels**: Each channel has an operation menu and it will pop up after pressing CH1 or CH2 button. **Channel coupling** To use Channel 1 as an example, input a sine wave signal with DC shift. Press CH1→Coupling→ AC to set "AC" coupling. It will pass AC component (Figure 2).

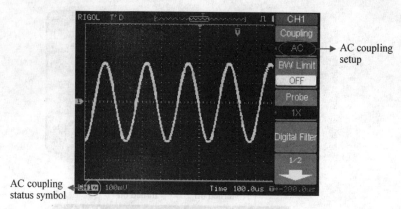

Figure 2 AC coupling setting
From http://www.rigol.com/download

Press CH1→Coupling→DC, to set DC coupling. It will pass both AC and DC components of the input signal. Press CH1→Coupling→GND, to set "GND" coupling, it disconnects the input signal (figures not shown).

2) **Volts/Div settings**: The Volts/Div control has **Coarse** or **Fine** configuration. The Vertical Sensitivity is 2 mV/div ~ 10 V/div. **Coarse**: It is the default setting of Volts/Div in a 1-2-5-step sequence from 2 mV/div, 5 mV/div, 10 mV/div, 20 mV/div... 10 V/div. **Fine**: This setting changes the vertical scale to small steps between the coarse settings. It will be helpful to adjust the waveform in smooth steps.

(4) **To Set up the Horizontal System**

The oscilloscope shows the time per division in the scale readout. Since all active waveforms use the same time base, the oscilloscope only displays one value for all the active channels, except when using Delayed Scan, or Alternative Trigger. The horizontal controls can change the horizontal scale and position of waveforms. The horizontal center of the screen is the time reference for waveforms. Changing the horizontal scale causes the waveform to expand or contract about the screen center. Horizontal position changes the displayed waveform position, relative to the trigger point.

1) **The Horizontal Knobs and Horizontal Menu**: The horizontal **position** knob adjusts the horizontal position of all channel (include Math) waveforms. The resolution of this control varies with the time base. Pressing this button clears trigger offset and moves the trigger point to the horizontal center of the screen. Use **scale** to select the horizontal time/div (scale factor) for the main or the

Delayed Scan time base. When Delayed Scan is enabled, it changes the width of the window zone by changing the Delayed Scan.

Horizontal Menu: Press the horizontal MENU button to display the horizontal menu.

2) **X-Y Format**: This format is useful for studying phase relationships between two signals. Channel 1 in the horizontal axis (X) and channel 2 in the vertical axis (Y), the oscilloscope uses a none-trigger acquisition mode, data is displayed as Figure 3.

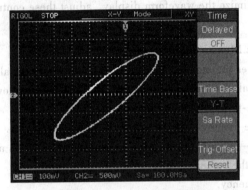

Figure 3   Lissajous curve
From http://www.rigol.com/download

【Procedure】

1. A Simple Measurement by Oscilloscope

To measure various signal accurately, we use a comparative measurement method in which we compare unknown signal with known signal, so we can measure voltages and frequencies of various waveform. The known signal used in this experiment is oscilloscope's own standard signal-square wave ($V_{p-p} = 2.0$ V, $f = 1$ kHz).

The method measuring voltage, period, and frequency by oscilloscope is as follows:

(1) **The measurement of alternative voltage**

If V/cm is set at "0.2 V/cm", and Y amplitude of signal is 4.5 cm, we can obtain peak-peak value of voltage which is

$$V_{p-p} = 2.0 \text{ V/cm} \times 4.5 \text{ cm} = 0.90 \text{ V}$$

(2) **The measurement of period and frequency**

If "T/cm" is set at "0.5 μs/cm", and the wavelength of signal is 5.6 cm, we can obtain period and frequency of this signal

$$T = 0.5 \ \mu s/cm \times 5.6 \ cm = 2.8 \ \mu s, \ f = 1/T = 3.6 \times 10^2 \ kHz$$

To quickly display a signal, please do the steps as follow:
1) Set the probe and the channel attenuations to 10X.
2) Connect signal to CH1 with the probe.
3) Press the AUTO button.

The oscilloscope sets the vertical, horizontal, and trigger controls at the best status automatically. To optimize the waveform display, adjust these controls manually to meet the requirements.

Selecting Automatic Measurements:
The oscilloscope takes automatic measurements on most signals. To measure the frequency and the peak-to-peak amplitude, do these steps as follows:
1) Measure peak-to-peak amplitude.
   Press Measure→Source→CH1 to set measurement source.
   Press Voltage→V_{p-p} to select peak-to-peak measurements and the result will be displayed on the screen.
2) Measure frequency.
   Press Measure→Source→CH1 to set measurement source.
   Press Time→Freq to select frequency measurements and the result will be displayed on the screen.

2. To observe Lissajous Curve

Lissajous figures

Most oscilloscopes have both an X and Y axis input. This is convenient for direct comparison of frequencies. When they have the same frequency and amplitude, the oscilloscope's beam should orbit around a perfect circle or ellipse on the screen. If the relative phase is just right, you may see a tilted straight line. If the two frequencies are slightly different, the pattern will not "stand still". If the frequencies are in exactly integral ratio, you'll get stable, pretty patterns called Lissajous figures(Figure 4). The ratio of frequencies can be "read off" the screen by counting "loops" of the pattern on the vertical and horizontal axes. Do this.

X-Y Format

This format is useful for studying phase relationships between two signals. Channel 1 in the horizontal axis (X) and channel 2 in the vertical axis(Y), the oscilloscope uses a none-trigger acquisition mode. To view the Lissajous curve, do these steps:

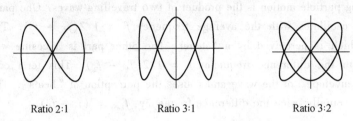

Ratio 2:1   Ratio 3:1   Ratio 3:2

Figure 4 Lissajous figures

(1) Press the AUTO button.
(2) Adjust the vertical knob to display approximately the same amplitude signals on each channel.
(3) Press the MENU in horizontal control area to display the menu.
(4) Press the Time Base soft button to select X-Y.
(5) Adjust the vertical and knobs to a desirable waveform display.

Note: In Y-T format, all sample rates are available. But in X-Y format, 100 MSa/s is not available. In common, decreasing the sample rate can display the waveform.

If the frequency ratio of two perpendicular harmonic vibrations is integral number, a stable wave form-Lissajous' figures will display on the screen. The simple relationship between Lissajous' figures and frequency ratio is as follows:

The number of tangent points of Lissajous' figures along X-axis Nx/the number of tangent points of the figure along Y-axis $N_y = f_y/f_x$. If we know one of the two frequencies f_x, f_y, according to frequency ratio determined by the number of tangent points in Lissajous' figures, another unknown frequency can be calculated.

3. To Observe Beat

Beat is a periodic variation in amplitude due to two waves of slightly different frequency. It arises from interference. Number of beats per second = Frequency difference.

E. g. Two sine waves with different frequencies

Two waves of equal amplitude, different frequencies and wavelengths are travelling in the same direction with the same wave speed.

$$y(x,t) = y_m \sin(k_1 x - \omega_1 t) + y_m \sin(k_2 x - \omega_2 t)$$
$$= 2y_m \cos\left[\frac{(k_1 - k_2)}{2}x - \frac{\omega_1 - \omega_2}{2}t\right] \sin\left[\frac{(k_1 + k_2)}{2}x - \frac{(\omega_1 - \omega_2)}{2}t\right]$$

This resulting particle motion is the product of **two** travelling waves. One part is a sine wave which oscillates with the average frequency $f = 1/2(f_1 + f_2)$. This is the frequency which is perceived by a listener. The other part is a cosine wave which oscillates with the difference frequency $f = 1/2(f_1 - f_2)$. This term controls the amplitude "envelope" of the wave and causes the perception of "beats". The "beat" frequency is actually twice the difference frequency, $f_{beat} = (f_1 - f_2)$.

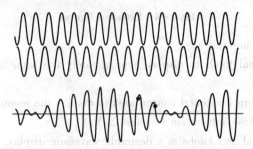

Figure 5 The Beat due to two waves of slightly different frequency

The "beat" wave oscillates with the average frequency, and its amplitude envelope varies according to the difference frequency.

Adjust the Generator to output two sine waves of similar frequencies. Input these two waves into channel CH1 and channel CH2. In oscilloscope, the mathematic functions include "add", "subtract", "multiply" and "FFT" for Channel 1 and Channel 2. The mathematic result can be measured by grid and cursor. Define CH1 or CH2 as source A, then define CH1 or CH2 as source B. Add source A and B. After finishing the above settings, the beat of wave will be shown on the screen.

[Data and Analysis]

1. A Simple Measurement

Waveform	(V) "V/cm"	Y/cm	V_{p-p}/V	(ms) "T/cm"	L/cm	T/ms	f/kHz	Signal generator's f/kHz
Sine wave								
Square wave								
Ramp wave								

2. Lissajous Figures

Plot the Lissajous figures you observed.

In your Lissajous curve, the ratio of two vibrating frequencies is _____.

3. Beat

Try to observe superposition of the two sine waves with slightly different frequencies traveling in the same direction-the "beat" wave. The frequency of beat you observed is _____.

[Questions]

1. What physical quantity can be measured directly by the oscilloscope? How many significant figures may be read out? What is the advantage in the measurement of the oscilloscope?

2. Can you observe non-periodical signals by the oscilloscope in this experiment?

3.5 Magnetism

3.5.1 Magnetic Force Due to a Current-carrying Wire

【Objective】

Use the magnetic force apparatus to verify that the magnetic force due to a current-carrying wire immersed in a perpendicular uniform magnetic field is proportional to **each** of the following parameters: length of the wire, electrical current flowing in the wire and magnitude of the magnetic field.

【Equipment】

The triple-beam balance, the variable current source, permanent magnet assembly with six horseshoe magnets, the power supply's digital ammeter, Vernier caliper, Lab stand, six interchangeable circuit board wire loops, current Balance Accessory Unit.

【Principle】

If a charged particle moves with some velocity, $\vec{v}$, through a uniform magnetic field, $\vec{B}$, it experiences a magnetic force given by

$$\vec{F} = q\vec{v} \times \vec{B} \tag{1}$$

where q is the charge of the particle. If the angle between the particle's velocity vector and the direction of the magnetic field is θ, the magnitude of the magnetic force may be rewritten as

$$F_B = qvB\sin\theta \tag{2}$$

The direction of the magnetic force vector may then be found with the familiar **right-hand rule**. Notice that the magnitude of the force is a maximum when $\vec{v} \perp \vec{B}$ and is identically zero when $\vec{v} \parallel \vec{B}$. See Figure 1.

Figure 2 shows a segment of wire carrying a current I that is located within a uniform magnetic field, $\vec{B}$. The force on each charged particle is given by the drift velocity of the charged particles. The volume of the wire that exists within the magnetic field is AL, where A is the wire's cross-sectional area and L is the length of wire that is **embedded within the magnetic field**. If we define n to be the number of charged particles per unit

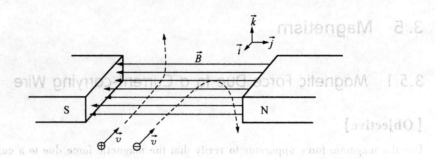

Figure 1 Two charged particles travel with some velocity, $\vec{v}$, through a uniform magnetic field, $\vec{B}$. As the charges pass through the magnetic field, each experiences a magnetic force, $\vec{F} = q\vec{v} \times \vec{B}$, due to their velocity, the direction and strength of the magnetic field and their charge, q. Note that here the positive charge experiences an upward magnetic force and the negative charge experiences a downward force

volume, at any instant there are nAL charges within that segment of wire. Therefore, we can write the magnetic force on a wire of length L as $\vec{F} = nALq\vec{v} \times \vec{B}$. Since the current flowing in a conductor is given as $I = nqvA$, the above equation becomes

$$\vec{F} = I\vec{L} \times \vec{B} \quad (3)$$

where $\vec{L}$ is the vector length of wire that points in the direction of the current I. Note that the direction of the current is defined as the direction in which positive charges move.

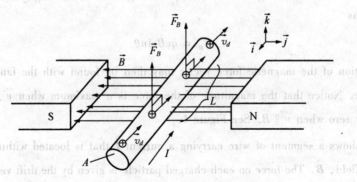

Figure 2 A representation of wire carrying a current I located within a uniform magnetic field, $\vec{B}$

3 CLASSICAL EXPERIMENTS

Our experimental setup is shown in Figure 3 and is described as follows. A permanent magnet assembly, comprised of six removable horseshoe magnets, is placed on a triple-beam balance, and the balance is then zeroed. A variable current source is connected to the current balance assembly, which has at one end a removable wire loop etched onto a circuit board. This wire loop is then placed into the permanent magnet assembly so the wire loop is perpendicular to the magnetic field but is not touching the magnets. Then, when a current flows through the wire loop, a magnetic force is created. Since the wire loop is stationary the magnetic force acts on the permanent magnet assembly causing its weight to either increase or decrease depending on the direction of the current and the orientation of the magnetic field. The **change** in the magnet assembly's weight is due to the magnetic force given by Equation (3).

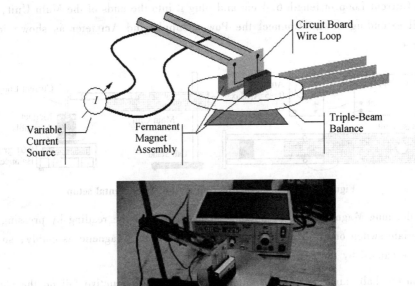

Figure 3 The experimental setup. A magnetic force is created when a current passes through the circuit board wire loop. This force acts on the permanent magnet assembly causing a change in its weight. The change in the magnet assembly's weight is directly proportional to the magnetic force

Three parameters may be altered in this experiment, and they are as follows:
(1) The **length of wire** may be varied by exchanging one wire loop for another.

(2) The **current amplitude** may be varied by adjusting the output from the power supply. (The direction of the current flow may also be altered.)

(3) The **strength of the magnetic field** may be altered by varying the number of horseshoe magnets in the magnet assembly. (The direction of the magnetic field may also be altered.)

In the present experiment we will vary three of the variables in the above equation: the current, the length of the wire, and the angle between the wire and the magnetic field. The force on the wire will be measured with an electronic balance.

【Procedure】

1. Force Versus Current

Use a Current Loop of length 6.4 cm and plug it into the ends of the Main Unit, with the foil extending down. Connect the Power Supply and Ammeter as shown in the Figure 4.

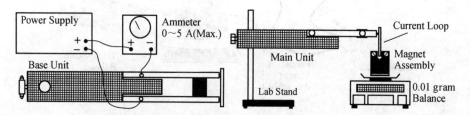

Figure 4 A schematic diagram of the experimental setup

Place the long Magnet assembly on the balance and tare the reading by pressing the appropriate switch on it. This subtracts the weight of the Magnetic assembly, so only the force caused by the current will be measured.

Position the Lab stand so the horizontal portion of the conductive foil on the Current Loop passes through the pole region of the magnets as shown in Figure 4. The Current Loop should not touch the magnets. Notice the right-most wire loop is printed on the **front and back** of the circuit board, effectively doubling the length of the wire shown on one side. **Be careful when removing and inserting these somewhat fragile circuit boards.**

Turn the current on. If the reading is negative, reverse the leads where they plug into the arms of the Main Unit. The measured weight is directly proportional to the force caused by the current moving through the magnetic field created by the Magnet Assembly.

Set the current to 0.5 A. Determine the new "mass" and calculate the corresponding force in Newtons. Record their values. Increase the current in 0.5 A increments to a maximum of 3.0 A, each time recording the corresponding "mass" and force readings.

Plot a graph of Force (vertical axis) versus Current (horizontal axis).

2. Force Versus Length of Wire

Start with the Power Supply turned off and a Current Loop of length 1.2 cm plugged into the ends of the Main Unit. Place the long Magnet assembly on the balance and take the reading by pressing the appropriate switch on it. Turn the Power Supply on, and adjust the current to 3.0 A. Insert the Current Loop into the pole region of the magnets. Record the "mass"/force readings. Turn the power supply off.

Repeat with Current Loops of length 2.2 cm, 3.2 cm, 4.2 cm, 6.4 cm and 8.4 cm. Each time record the corresponding mass/force reading.

Plot a graph of Force (vertical axes) versus Length (horizontal axis).

3. Force Versus Angle

(1) Set up the apparatus as shown in Figure 5.
(2) Place the short Magnet assembly on the balance. With no current flowing, adjusting the reading to 0.00 g. Set the angle to 0° with the direction of the coil of wire approximately parallel to the magnetic field.
(3) Set the current to 2.5 A. Record the "mass"/force value. Increase the angle in 10° increments up to 90°, and then in −10° increments to −90°. At each angle, repeat the mass/force measurement.
(4) Plot a graph of Force (vertical axis) versus Angle (horizontal axis).

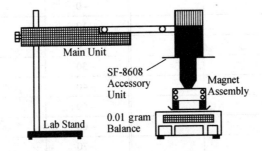

Figure 5 Setup about force vs. angle

[Data and Analysis]

Table 1 Force versus Current

Current I/A	0.0	0.5	1.0	1.5	2.0	2.5	3.0
Mass m/g							
Force F/N							

Note: The mass reading is proportional to the actual force, which is given by the equation $F = mg$. While doing the experiment, try to find what is the nature of the relationship between these two variables?

Table 2 Force versus Length of Wire

Length L/cm	1.2	2.2	3.2	4.2	6.4	8.4
Mass m/g						
Force F/N						

Again, while doing the experiment try to find what is the nature of the relationship between these two variables?

Table 3 Force versus Angle (Set the current to 2.5 A)

Angle/(°)	Mass/g	Force/N	Angle/(°)	Mass/g	Force/N
0			-0		
10			-10		
20			-20		
30			-30		
40			-40		
50			-50		
60			-60		
70			-70		
80			-80		
90			-90		

3   CLASSICAL EXPERIMENTS   187

Think on the following questions: What is the relationship between these two variables? How do changes in the angle between the current and the magnetic field affect the force acting between them? What angle produces the greatest force? What angle produces the least force?

Show that it is only the **horizontal** portion of the circuit board wire that contributes to the vertical magnetic force. In other words, show that the vertical portion of the wire does not vary the weight of the magnet assembly.

[Caution]
(1) Do not touch the metal arms of the circuit board holder while current is flowing through them!
(2) Be careful with the small etched circuit boards when inserting and removing them—they can break easily!
(3) Keep the current **below 5 A** throughout the experiment!
(4) The power supply should be set to **constant current mode.** To do so, turn the DC VOLTAGE ADJUST knob fully clockwise, then adjust the DC CURRENT ADJUST knob to obtain the desired output current.

[Questions]
1. How will you verify that the magnetic force is proportional to each parameter?

2. Is it important that the triple-beam balance be properly zeroed before the experiment begins? Why or why not?

3. How will you insert the circuit-board wire loops into the permanent magnet assembly? Is the orientation and position of the wire relative to the magnets important?

4. How did you measure the length of the wire? Did you measure the entire length of the conductor?

5. Is this experimental setup sensitive enough to measure the earth's magnetic field? If not, what can be done to make this measurement possible?

3.5.2 Magnetic Field Measured by Hall Effect

Magnetic field is an important physical quantity. Its effect on the matter is extensively applied in modern technology, such as magnetic recording, magnetic resonance, and biological effect of magnetic field, etc. Many methods have been used to measure magnetic field, such as magnetic resistance, magnetic induction, nuclear magnetic resonance (NMR) and Hall effect.

【Objective】
- To know the fundamental mechanism of measuring magnetic field by Hall effect.
- To learn how to avoid negative effect by using symmetrical measurement.

【Equipment】
Magnetic field meters.

【Principle】

1. Hall Effect

When a thin semiconductor board through which a current is passing is subjected to a magnetic field, and the plane of this board is perpendicular to external magnetic field in the Figure 1, carriers will be aggregated to one side of thin board under the action of Lorenz force, inducing in potential difference perpendicular to the direction of current. This effect is called Hall effect, named after its discoverer, E. H. Hall in 1879. The potential difference, U_H, built during this process is called Hall potential difference with magnitude of millivolts.

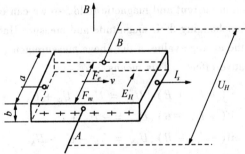

Figure 1 The diagram of principle of Hall effect

Experiment showed that Hall potential difference could be written as follows,

$$U_H = k \frac{BI_s}{b} = K_n BI_s$$

Where, K_n is the sensitivity of Hall device (in this experiment, Hall device means thin semiconductor board), B is the magnitude of magnetic field, I_s is the current passing through Hall device, b is the thickness of Hall device.

If the sensitivity of Hall device K_n and current passing through this device I_s are known, and Hall potential difference U_H is measured by this experiment, we can calculate the intensity of magnetic induction B indirectly.

2. Negative Effect of Hall Effect

When Hall potential difference is measured in this experiment, several negative effects will affect the value of measurement.

(1) **Nonequivalent potential difference**: When a Hall device is fabricating, it is too difficult to make electrodes A and B at equipotential line. Under this condition, when current I_s passes through Hall device, a potential U_0 is built even if no external magnetic field is added. Fortunately, U_0 is independent of external magnetic field, and it only depends on the magnitude and the direction of the current I_s.

(2) **Nernst effect**: Because the contact resistance between every electrode and hall device is generally different, so is Joule heat of current at contact point. A potential difference from temperature difference between the two electrodes is therefore built, inducing in current of temperature difference. The temperature dependent current is curved in the magnetic field, so that an additive potential difference U_N is built in the direction of Hall potential difference U_H.

The additive potential difference induced by all these negative effects depends on the magnitude and direction of current and magnetic field, so we can change their direction under the condition of unchanging their magnitude and measure Hall potential difference many times, then get the average value of the above measurement, in order to avoid the influence of these negative effects. This means

$$\text{if}(+I_s, +B), U_{H1} = U_H + U_0 + U_N$$
$$\text{if}(+I_s, -B), U_{H2} = -U_H + U_0 - U_N$$
$$\text{if}(-I_s, -B), U_{H3} = U_H - U_0 - U_N$$
$$\text{if}(-I_s, +B), U_{H4} = -U_H - U_0 - U_N$$

Therefore, $U_H = \dfrac{1}{4}(U_{H1} - U_{H2} + U_{H3} - U_{H4})$

3. Distribution of Magnetic Field in Long Straight Solenoid

According to theory of electromagnetism, distribution of magnetic field in solenoid can be concluded as follows,

$$B = \frac{1}{2}\mu_0 N I_M (\cos\beta_1 - \cos\beta_2)$$

where, N is the number of turns per unit length, I_M is excitation current.

For inner of long straight solenoid, $\beta_1 \to 0, \beta_2 \to \pi$, so

$$B = \mu_0 N I_M \tag{1}$$

Near one end of long straight solenoid, $\beta_1 \to \frac{\pi}{2}, \beta_2 \to \pi$, so

$$B = \frac{1}{2}\mu_0 N I_M \tag{2}$$

In this experiment, students should use Hall device to measure the intensity of magnetic induction at different point from the central to one end of solenoid, and compare the values of measurement to those of theoretical calculation.

[Procedure]

(1) Connect the wires according to the instruction of equipment. Do not start the experiment before the teacher check it.

Attention: do not misconnect the wires for both work current of Hall device and excitation current of solenoid, otherwise, it will damage the equipment.

(2) Turn off double-pole-double-throw switch for I_s, I_M turn on double-pole-double-throw for Hall potential difference U_H by upper throw, display value of U_H should be "0.00", if there is a shift, adjust the "zero" resistor on the board to calibrate.

(3) Adjust work current (I_s) of Hall device of magnetic field meter up to 10.00 mA, and excitation current (I_M) of solenoid up to 1.00 A.

(4) Move the transfer probe for Hall device slowly, put Hall device to one end of solenoid ($x = 0$), turn on switches for I_s and I_M by upper throw (the state is called ($+I_s, +B$)), you can measure Hall potential difference U_{H1} at $x = 0$ under the condition of ($+I_s, +B$); invert magnetic field (turn on switch for I_M by down throw), you can measure Hall potential difference U_{H2} at $x = 0$ under the condition of ($+I_s, -B$); invert work current I_s, you can measure Hall potential difference U_{H3} at $x = 0$ under the condition of ($-I_s, -B$); at last, invert magnetic field and measure Hall potential difference U_{H4} at $x = 0$ under the condition of ($-I_s, +B$).

According to the results measured above, we can get

$$U_H = \frac{1}{4}(U_{H1} - U_{H2} + U_{H3} - U_{H4})$$

$$= \frac{1}{4}(|U_{H1}| + |U_{H2}| + |U_{H3}| + |U_{H4}|)$$

The corresponding intensity of magnetic induction is

$$B_0 = \frac{U_H}{K_n I_s} \qquad (3)$$

(5) Change the position of Hall device in the solenoid, and take account into coordinates given in data table, measure Hall potential difference ($U_{H1}, U_{H2}, U_{H3}, U_{H4}$) at different position in the solenoid, till Hall device is moved to central point ($x = 15$ cm) in it. Then calculate intensity of magnetic induction at each point according to formula (3).

(6) From theoretical formula (1) and (2), calculate intensity of magnetic induction both central point and one end of solenoid and compare them with the results of experiment.

(7) Central point of solenoid is set as origin of coordinate axes, x as x-axis, B_x/B_{15} as y-axis (where, B_{15} means measurement value of magnetic field at $x = 15$ cm), draw a $B_x/B_{15} - x$ curve on coordinate paper.

[Attention]

(1) The switch for U_H keeps connecting all the time during the experiment, otherwise, the display for U_H will be "1".

(2) When you record the data or change x (the position of Hall device), please turn off the switch for I_M to avoid electric heating solenoid and influencing the results of measurement.

(3) Hall device is fragile, please do not pump or press it, you should move it slowly when changing its position.

[Data and Analysis]

1. Experimental parameters

Work current $I_s = 10.00$ mA, the number of turns per unit length $N =$ _____ turns/m
The intensity of Hall device $K_n =$ _____ mV/(mA · T), excitation current $I_M = 1.00$ A

2. Experimental data sheet

Position of Hall device/cm	U_{H1}/mV	U_{H2}/mV	U_{H3}/mV	U_{H4}/mV	U_{H}/mV	B/T
0.00						
1.00						
2.00						
3.00						
6.00						
9.00						
12.00						
15.00 (Central point)						

3. Calculate theoretical value

$$B_{15} = \mu_0 N I_M ; \quad B_0 = \frac{1}{2}\mu_0 N I_M =$$

4. Draw a $B_x/B_{15} - x$ curve (x is used as x-axis, B_x/B_{15} as y-axis)

2. Experimental data sheet

Position in Hall device/cm	I_S/mA	V_H/mV	U_{H1}/mV	U_{H2}/mV	U_{H3}/mV	U_{H4}/mV	V_H/mV	B_Z/T
0.00								
1.00								
2.00								
3.00								
6.00								
9.00								
12.00								
15.00 (Central point)								

3. Calculate theoretical value

$$B_0 = \mu_0 N I_S / L = \frac{1}{2}\mu_0 N I$$

4. Draw a $B_Z/B_0 - x$ curve (x is used as x-axis, B_Z/B_0 as y-axis).

3.5.3 Transformer

【Objective】
- To study the operation of a transformer.
- To examine the effect of core configuration on the output voltage gain.
- To compare between step-up and step-down transformers.
- To compare between no-load and full-load operation.

【Equipment】

The PASCO SF-8616 Basic Coils Set.
Low voltage ac power supply 0-6 VAC, 0-1 amp such as PASCO Model SF-9582.
Resistance box.
Banana connecting leads for electrical connections.
Multimeter(2).

【Principle】

In electric power distribution systems, it is desirable at both the generating end (the electric power plant) and the receiving end (home or factory) to deal with relatively low voltages. For example, no one wants a child's train to operate at, say, 10 kV.

On the other hand, in the transmission of electric energy from the generating plant to the consumer, we want the lowest possible current (and thus the largest possible potential difference) so as to minimize the i^2R ohmic losses in the transmission line. Therefore, there is a fundamental mismatch between the requirements for efficient transmission on the one hand and safe generation and consumption on the other hand. We need a device that can, as design considerations require, raise or lower the potential difference in a circuit, keeping the product IV essentially constant.

The transformer solves the problem (see Figure 1). It consists of two coils wound around a soft iron coil. One is referred to as the primary coil, with N_1 turns, and the other is the secondary coil, with N_2 turns. The primary coil is connected to an alternating-current (AC) source with electromotive force (E. M. F.), $\varepsilon(t)$, and the secondary is connected to a resistive load via a switch.

According to Faraday's Law, an alternating current in the primary coil induces a self alternating magnetic flux $\Phi_B(t)$, such that

$$\varepsilon(t) = V_1 = - N_1 d\Phi_B/dt \qquad (1)$$

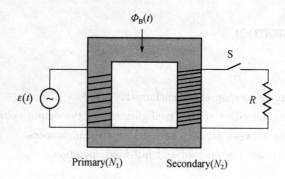

Figure 1 The basic transformer

where, V_1 is the potential difference developed across the primary coil. The magnetic flux is linked to the secondary coil through the iron core. Thus, the magnetic flux rate is the same for both coils, therefore, using Eq. (1), we get

$$- d\Phi_B/dt = V_1/N_1 = V_2/N_2 \qquad (2)$$

where V_2 is the potential difference developed across the secondary coil. Now, if $N_2 > N_1$, we speak of a step-up transformer; and if $N_2 < N_1$, we speak of a step-down transformer. The voltage gain, G, may be defined as the ratio of the output, to the input voltages.

When the switch is open (no load operation), no current exists in the secondary coil and therefore, no power is delivered to the transformer, and the primary coil acts as a pure inductance. Whereas, if the switch is closed, a current i_2, is set through the secondary coil, and the two windings appear to be as a fully coupled mutual inductance.

Actually, the closed switch operation is rather complex to analyze. Therefore we take advantage of the overall view provided by the conservation of energy principle. For an ideal transformer with a resistive load this tells us that

$$P_{in} = P_{out} \qquad (3)$$

Where P_{in}, and P_{out} are the input and output powers. Transformer efficiency may be expressed as

$$\text{Efficiency} = P_{out}/P_{in} \times 100\% \qquad (4)$$

In practice no transformer is of 100% efficiency due to power losses. Some of the main reasons for these losses are: the resistance of the coils, the magnetic leakage, and the hysteresis losses (due to magnetization properties of the core).

[Procedure]

1. Core Configuration

(1) Set up the two coils labeled 400-turn as shown in Figure 2 (no core is used).
(2) Set the voltage of the supply to 6 V.
(3) Measure input and output voltages, calculate the voltage gain and record in Table 1.
(4) Repeat step (3), changing the core configuration as shown in Figure 3.

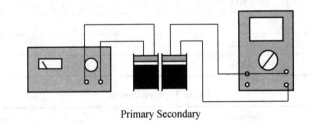

Primary Secondary

Figure 2   Input and output voltage measurements

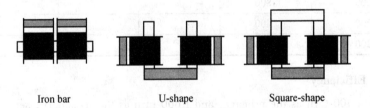

Iron bar   U-shape   Square-shape

Figure 3   Core configurations

2. Step-up vs Step-down Transformer

(1) Using the core configuration that gave the maximum voltage gain in part 1, set up the coils as shown in Figure 2.
(2) Measure input and output voltages, calculate the voltage gain and record in Table 1.
(3) Fixing the primary coil to 400-turns, repeat step (2), changing the secondary coil according to Table 2.
(4) Use the 3 200-turn coil as your primary, repeat step (2), changing the secondary coil as given in Table 3. Record the data.

Table 1 Both coils of 400-turns

Core	V_{in}/V	V_{out}/V	voltage gain $G = V_{out}/V_{in}$
No core			
Iron bar			
U-shape			
Square-shape			

Table 2 Step-up

Primary	Secondary	V_{in}/V	V_{out}/V	voltage gain G
400	400			
400	800			
400	1600			
400	3200			

Table 3 Step-down

Primary	Secondary	V_{in}/V	V_{out}/V	voltage gain G
3200	1600			
3200	800			
3200	400			
3200	200			

3. Power Efficiency

(1) Using a 400-turn as the primary, and 1 600-turn as the secondary, set up the coils as shown in Figure 4, using the square core configuration.

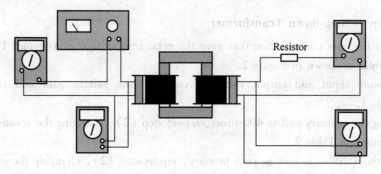

Figure 4 Current and voltage measurements

(2) Set the ammeters to milli-Amp range. Set the resistor to 1 000 Ω.
(3) Measure the input and output voltages and currents. Record in Table 4.

Table 4 Voltages and currents

R/Ω	V_{in}/V	I_{in}/mA	V_{out}/V	I_{out}/mA
1000				
1100				
1200				
1300				
1400				
1500				

(4) Calculate input and output powers, voltage gain, and the power efficiency. Record in Table 5.
(5) Repeat steps (3) and (4) for resistor values given in the table.

Table 5 power efficiency

R/Ω	P_{in}/W	P_{out}/W	voltage gain G	Efficiency
1000				
1100				
1200				
1300				
1400				
1500				

[Questions]

1. Explain why stepping-up the voltage in the beginning of the transmission lines, reduces the ohmic loses.

2. Which core configuration gives the maximum voltage gain? Explain.

3. When no load is connected to the secondary coil, does the transformer dissipate energy? Why?

4. Is the power efficiency, that you got, equals 100%? If not, can you explain the reason?

3.5.4 Faraday's Law of Induction

【Objective】

To demonstrate the Faraday's Law of electromagnetic induction.

【Equipment】

One large and two small (with handles) coils, plastic triangles, T-base BNC connector.

【Principle】

In 1830 Joseph Henry and Faraday observed that an E. M. F. (*electromotive force*) is setup in a coil placed in a magnetic field whenever the flux through the coil changes. This effect is called Electromagnetic Induction. If the coil forms a part of a close circuit, the E. M. F. causes a current to flow in the circuit. E. M. F. setup in the coil is called "induced E. M. F" and the current thus produced is termed as "Induced Current".

Any change in the magnetic environment of a coil of wire will cause a voltage (E. M. F.) to be "induced" in the coil. No matter how the change is produced, the voltage will be generated. The change could be produced by changing the magnetic field strength, moving a magnet toward or away from the coil(Figure 1,2), moving the coil into or out of the magnetic field, rotating the coil relative to the magnet, etc. Experiments show that the magnitude of E. M. F. depends on the rate at which the flux through the coil changes. It also depends on the number of turns on the coil.

There are various ways to change magnetic flux of a coil such as:
(1) By changing the relative position of the coil with respect to a magnet.
(2) By changing current in the coil itself.
(3) By changing current in the neighboring coil.
(4) By changing area of a coil placed in the magnetic field, etc.

Faraday's law of electromagnetic induction: Faraday was the first scientist who performed a number of experiments to discover the facts and figures of electromagnetic induction, he formulated the following law:

The EMF generated is proportional to the rate of change of the magnetic flux.

When magnetic flux changes through a circuit, an E. M. F is induced in it which lasts only as long as the change in the magnetic flux through the circuit continues. Quantitatively, induced E. M. F is directly proportional to the rate of change of magnetic flux through the coil. i. e. Average emf

$$\varepsilon = - N\Delta\Phi/\Delta t$$

Where N is the number of turns in the coil, magnetic flux $\Phi = \int \vec{B} \cdot \vec{A} = B \cdot A\cos\theta$ for a magnetic field (B) which is constant over the area (A). Here θ is the angle between the magnetic field vector and the magnitude of the area vector A.

The negative sign indicates that the induced current is such that the magnetic field due to it opposes the magnetic flux producing it. Consider a bar magnet and a coil of wire. When the N-pole of magnet is approaching the face of the coil, it becomes a north face by the induction of current in anticlockwise direction to oppose forward motion of the magnet.

Figure 1   Moving a magnet toward the coil

When the N-pole of the magnet is receding the face of the coil becomes a south pole due to a clockwise induced current to oppose the backward motion.

Figure 2   Moving a magnet away from the coil

In this experiment we will use a large (field) coil with $N = 200$ turns of radius rs (approx.) = 10.5 cm to create magnetic field, and two search coils: one with $N = 400$ turns and another with $N = 2\ 000$ turns, both of radius rs (approx.) = 3 cm, to measure the induced *emf*. During the experiment, the area of the field and search coils is constant.

The current in the field coil, however, oscillates thus creating an oscillatory magnetic field (flux) through the search coil positioned at the center of the field coil. As a result, an average emf is induced in the search coil, given by $\varepsilon = - NA\Delta B/\Delta t$, where $\Delta B/\Delta t$ is the time rate of change (oscillation) of the magnetic field, N and A are the number of turns and area of the search coil, respectively. When current, i, runs through a circular coil with N turns each of radius, r, a magnetic field is created. Its strength, B, at the center of the coil, is given by the Biot-Savart Law: $B = N\mu i/2r$, where $\mu = 4\pi \times 10^{-7} T$ is the so called permeability constant.

[Procedure]
(1) Set the Function Generator to produce a triangular wave with amplitude of 5 V and frequency of 2 kHz.
(2) Build the circuit shown in Figure 3.

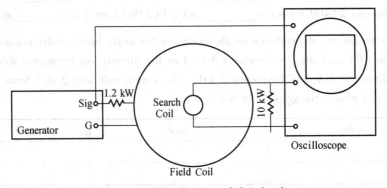

Figure 3   The diagram of the circuit

Note: the primary side of the circuit includes a 1.2 kW resistor, so that the current in the field coil can be calculated from the voltage delivered by the function generator. The resistance and self-inductance of the field coil are small compared to those of the 1.2 kW resistor. The 10 kW resistor in the secondary (search) coil damps out some of the unwanted oscillations (noise), so a relatively square wave emf can be seen on the oscilloscope screen. Position the triangular wave/voltage (generating the oscillatory magnetic field) and the resulting square wave emf as shown in Figure 4 to facilitate data reading from the oscilloscope screen. Use both Inputs (#1 and #2) of the scope.

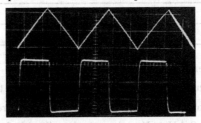

Figure 4   Triangular wave and the resulting square
wave on the oscilloscope screen

[Data and Analysis]

(1) Using the known values of the circuit components calculate the strength of the magnetic field at the center of the field coil with rs (approx.) = 10.5 cm.

B = _____ [T]

(2) Predict the induced *emf* for the search coils with N = 400 and N = 2 000 turns and radius (approx.) = 1.5 cm. Measure those emfs. Compare the predicted and measured values.

Predicted: E(400 turns) = _____ mV; E(2 000 turns) = _____ mV
Measured: E(400 turns) = _____ mV; E(2 000 turns) = _____ mV

(3) Measure the *emf* dependence on the cosine of the angle between the magnetic field vector (B) and the area vector (A). Use the already set triangular wave with amplitude of 5 V and frequency 2 kHz, the search coil with 2 000 turns and the provided plastic triangles. Plot the data.

θ/deg.	$\cos\theta$	emf/mV
0		
30		
45		
60		
90		

(4) Measure the *emf* dependence on the strength of the magnetic field/flux (at a fixed frequency of 2 kHz). For the purpose, vary the amplitude/voltage of the triangular wave provided by the function generator. Starting amplitude: 1.0 V; step: 1.0 V; end amplitude: 7.0 V. Use the search coil with 2 000 turns. Plot the emf voltage versus the generator (triangular wave) voltage.

Ampl./V	1.0	2.0	3.0	4.0	5.0	6.0	7.0	8.0
emf/mV								

(5) Measure the *emf* dependence on the rate of change of the magnetic field/flux (at a fixed generator amplitude of 5 V). For the purpose, vary the frequency of the triangular wave provided by the function generator. Starting frequency: 1 kHz; step: 1 kHz; end frequency: 8 kHz. Use the search coil with 2 000 turns. Plot the emf voltage versus the generator (triangular wave/magnetic field) frequency.

3 CLASSICAL EXPERIMENTS

f/kHz	1.0	2.0	3.0	4.0	5.0	6.0	7.0	8.0
emf/mV								

(6) Measure the emf intensity along an axis perpendicular to the field coil, at various distances from the plane of the coil, keeping the generator voltage/triangular wave amplitude (5 V) and frequency (2 kHz) constant. Plot the emf voltage versus the distance from the center of the field coil.

Distance/cm	0	5	10	15	20	25	30	35
emf/mV								

[Questions]

1. Predict the direction of the magnetic field for different locations around a bar magnet and an electromagnet.

2. Compare and contrast bar magnets and electromagnets.

3. Identify the characteristics of electromagnets that are variable and what effects each variable has on the magnetic field's strength and direction.

4. Relate magnetic field strength to distance quantitatively and qualitatively.

5. Identify equipment and conditions that produce induction.

6. Compare and contrast how both a light bulb and voltmeter can be used to show characteristics of the induced current.

7. Predict how the current will change when the conditions are varied.

8. Explain what happens when the magnet moves through the coil at different speeds and how this affects the brightness of the bulb and the magnitude and sign of the voltage.

9. Explain the difference between moving the magnet through the coil from the right side versus the left side, through the big coil versus the smaller coil.

4 INNOVATION EXPERIMENTS

4.1 Optical Tweezers

[Introduction]

Since Ashkin et al. published their seminal paper "Observation of a single-beam gradient force optical trap for dielectric particles" in 1986, the technique is now referred to as "optical tweezers" or "optical trapping".

Optical tweezers are capable of manipulating nano-meter and micrometer sized dielectric particles by exerting extremely small forces via a highly focused laser beam. The beam is typically focused by sending it through a microscope objective. The narrowest point of the focused beam, known as the **beam waist**, contains a very strong electric field gradient. It turns out that dielectric particles are attracted along the gradient to the region of strongest electric field, which is the center of the beam. The laser light also tends to apply a force on particles in the beam along the direction of beam propagation. It is easy to understand why if you imagine light to be a group of tiny particles, each impinging on the tiny dielectric particle in its path. This is known as the **scattering force** and results in the particle being displaced slightly downstream from the exact position of the beam waist.

In essence, optical tweezers rely upon the extremely high gradient in the electric field produced near the beam waist of a tightly focused laser beam, which creates a force sufficient to trap micron-sized dielectric particles in three dimensions. Using various techniques, these trapped particles can then be manipulated and forces on the objects in the trap can be measured. The forces that such an instrument is capable of measuring are of the order of one to 100 pico-Newtons (pN).

Optical traps are very sensitive instruments and are capable of the manipulation and detection of subnanometer displacements for sub-micron dielectric particles. For this reason, they are often used to manipulate and study single molecules by interacting with a bead that has been attached to that molecule. DNA and the proteins and enzymes that interact with it are commonly studied in this way.

[Objective]

Grasp the principle of optic tweezers and know how manipulate it.

[Principle]

When trying to understand the origin of the forces acting within optical tweezers, two distinct approaches may be adopted, one based on ray optics, and the other on the electric field associated with the light.

An optical tweezer is a scientific instrument that uses a focused laser beam to provide an attractive or repulsive force, depending on the index mismatch (typically on the order of pico-Newtons) to physically hold and move microscopic dielectric objects. Dielectric objects are attracted to the center of the beam, slightly above the beam waist. The force applied on the object depends linearly on its displacement from the trap center just as with a simple spring system.

Since a light beam carries a linear momentum of h/λ per photon, the refraction of light by a transparent object results in a change in photon momentum and a corresponding reaction force acting on the object, see Figure 1.

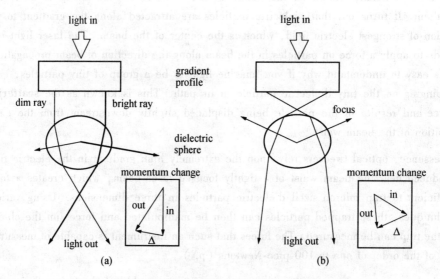

Figure 1 Momentum changes for light and dielectric ball

This approach can also allow for forces generated by reflection or scattering from the interface; however, as the trapped object is usually suspended in a fluid of similar refractive index, the resulting Fresnel reflections and corresponding recoil forces are small and in the main are ignored. Figure 2 shows the reflection and refraction of light rays at the surface of a dielectric sphere and the resulting forces acting upon it. And the counter-intuitive aspects of optical tweezers are immediately apparent.

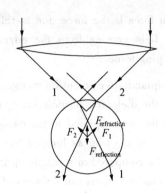

Figure 2   Schematic diagram showing the force on a dielectric sphere due to both reflection and refraction of two rays of light

(1) The intensity profile of the beam cross-section results in a force acting to move the object into the centre of the beam, that is the force arising from refraction of the light can be in the opposite sense to that from light scattering/reflection alone.

(2) If a beam incident from above is tightly focused, then it is possible to generate a force that acts to lift the object up towards the focus, thereby creating a three-dimensional trap with a single laser beam.

(3) Note that reversing the direction of a ray does not change the direction of the force; this illustrates nicely that the forces associated with refraction are linked to the beam intensity rather than to the direction of propagation.

Radiation pressure is a force per unit area on an object due to the change in momentum of light. All light consists of photons that each has momentum p. For light of specific wavelength λ, the magnitude of the momentum of a single photon is given by $p = h/\lambda$. The intensity of the light is determined by the number of photons passing through a given area per unit time. The momentum flux of photons from light of intensity given by the pointing vector S is

$$d(dP/dt) = (n/c)SdA$$

where n is the index of refraction, c is the speed of light, P is the total momentum of the photons, and dA is an element of area normal to S. Thus, in principle one can directly calculate the force on a given area due to the light momentum flux on that area. The total force on an object due to refraction of light is

$$F = (n/c)\iint (S_{in} - S_{out}) \, dA$$

The attractive force due to refraction of the light at the surface of the bead must be sufficient to overcome any other forces acting to push the bead out of the trap. One such force is the force due to reflection at the bead surface, illustrated in Figure 2. The light coming from the edges of the objective lens contributes the most to the trapping force.

For quantitative scientific measurements, most optical traps are operated in such a way that the dielectric particle rarely moves far from the trap center. The reason for this is that the force applied to the particle is linear with respect to its displacement from the center of the trap as long as the displacement is small. In this way, an optical trap can be compared to a simple spring, which follows Hooke's Law, $F = -kx$, where k is the elastic constant or stiffness of the optical trap. In the absence of any damping (i.e. in air or in vacuum) the result would be an oscillator with resonant frequency given by

$$f_{res} = \frac{1}{2\pi}\left(\frac{k}{m}\right)^{\frac{1}{2}}$$

Proper explanation of optical trapping behavior depends upon the size of the trapped particle relative to the wavelength of light used to trap it.

➤ In cases where the dimensions of the particle are greater than this wavelength, a simple ray optics treatment is sufficient.
➤ On the other hand, if the wavelength of light exceeds the particle dimensions, then the particles must be treated as tiny electric dipoles in an electric field.

Case1: Tweezers force: ray optics $a \gg \lambda$

For objects larger than the wavelength of the laser, the ray optics approach gives remarkably accurate estimates of the observed values of Q.

Case2: Tweezers force: electromagnetic field/ Rayleigh $a \ll \lambda$

For particles smaller than the wavelength of the laser beam, the ray optical approach is less satisfactory and it is better to consider the forces in terms of the electric field near the trapped particle. As before, the forces can be divided into those arising from scattering of the light and those arising from an intensity gradient. For an object of radius a and light beam of intensity I_0, the force resulting from the light scattering is,

$$F_{scat} = \frac{8\pi}{3}k^4 a^6 \frac{\sqrt{\epsilon_2}}{c}\left(\frac{\epsilon_1 - \epsilon_2}{\epsilon_1 + 2\epsilon_2}\right)^2 S$$

Radiation pressure:

where $\sqrt{\epsilon_1}$ is the refractive index of the object, $\sqrt{\epsilon_2}$ is the refractive index of the surrounding medium. Although not apparent from this form of the equation, the scattering force is directed perpendicular to the wavefronts of the incident light, that is, objects are pushed in the direction of light propagation.

The intensity gradient near the beam focus gives rise to a gradient force, which is equivalent to the refraction of the light rays, given by

$$F_{grad} = \frac{2\pi}{c} a^3 \left(\frac{\epsilon_1 - \epsilon_2}{\epsilon_1 + 2\epsilon_2} \right) \nabla |S|$$

Gradient force:

Explicit in this equation is that the force is directed towards the region of highest light intensity. For a full three-dimensional trap to be established, we require the gradient force to exceed the scattering force $F_{grad} >> F_{scat}$.

[**Experimental Design and Construction**]

The most basic optical tweezer setup (Figure 3) will likely include the following components:

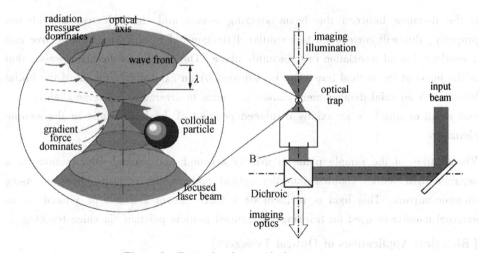

Figure 3 Example of an optical tweezer setup

Ashkin, J. M. Dziedzic, J. E. Bjorkholm, and S. Chu, Opt. Lett. 11, 288-290 (1986)

(1) a laser;
(2) a beam expander;
(3) some optics used to steer the beam location in the sample plane;

(4) a microscope objective and condenser to create the trap in the sample plane;
(5) a position detector (e. g. quadrant photodiode) to measure beam displacements;
(6) and a microscope illumination source coupled to a CCD camera.

The Nd:YAG laser (1 064 nm wavelength) is the most common laser choice because biological specimens are most transparent to laser wavelengths around 1 000 nm. This assures as low an absorption coefficient as possible, minimizing damage to the specimen, sometimes referred to as opticution.

Perhaps the most important consideration in optical tweezer design is the choice of the objective. A stable trap requires that the gradient force, which depends upon the numerical aperture (NA) of the objective, be greater than the scattering force. Suitable objectives typically have a NA between 1. 2 and 1. 4.

Expanding the beam emitted from the laser to fill the aperture of the objective will result in a tighter, diffraction-limited spot. While lateral translation of the trap relative to the sample can be accomplished by translation of the microscope slide, most tweezer setups have additional optics designed to translate the beam to give an extra degree of translational freedom.

If the distance between the beam steering lenses and the objective are chosen properly, this will correspond to a similar deflection before entering the objective and a resulting lateral translation in the sample plane. The position of the beam waist, that is the focus of the optical trap, can be adjusted by an axial displacement of the initial lens. Such an axial displacement causes the beam to diverge or converse slightly, the end result of which is an axially displaced position of the beam waist in the sample chamber.

Visualization of the sample plane is usually accomplished through illumination via a separate light source coupled into the optical path in the opposite direction using dichroic mirrors. This light is incident on a CCD camera and can be viewed on an external monitor or used for tracking the trapped particle position via video tracking.

[Biological Applications of Optical Tweezers]

1. Optical Tweezers in a"Landscape"——Cell Sorting

By using an electrical charge that the cell is trapped in, the cell are then sorted based on the fluorescence intensity measurements. The sorting process is undertaking by an electrostatic deflection system that diverts cell into containers based upon their charge. The main mechanism for sorting is the arrangement of the optical lattice points. As the cell flow through the optical lattice, there are forces due to the particles drag force that is competing directly with the optical gradient force from the optical lattice point. By shifting the arrangement of the optical lattice point, there is a preferred optical path where the optical forces are dominate and biased. With the aid of the flow of the cells, there is a resultant force that is directed along that preferred optical path. Hence, there is a relationship of the flow rate with the optical gradient force. Competition of the forces in the sorting environment need fine tuning to succeed in high efficient optical sorting. The need is mainly with regards to the balanced of the forces; drag force due to fluid flow and optical gradient force due to arrangement of intensity spot.

2. Observing Single Biological Motors at Work

A landmark in the study of single motor proteins came in 1993 when Svoboda et al. measured the individual steps taken by the molecular porter, kinesin, as it walked along a fixed microtubule track. In their experiments, a single kinesin molecule was bound to a plastic microsphere and this was then held close to its microtubule track, which had been fixed to a microscope coverslip. The bathing medium was a buffered salt solution containing the chemical fuel ATP. When the kinesin motor and microtubule track interacted, the bead was pulled along by the kinesin and the nanometre scale displacements and pico-Newton forces produced were measured. The crucial observation made, and indeed a testament to the incredible sensitivity of their method, was that they could identify discrete 8 nm steps taken by the kinesin molecule. Rather than moving smoothly like an ensemble would do, the single molecule moved in a stochastic jerky fashion. They found that the motor paused for a random interval after taking each step as it waited for a fresh ATP molecule to arrive.

Molecular-scale motions powered by just 1 ATP molecule (equivalent to less than one tenth of the energy of a single photon) were being observed in real time, with no signal averaging required. Figure 4 shows data from a study (N. J. Carter and R. A. Cross) using the same protein system.

In the experiment a single (double-headed) kinesin molecule was attached to a latex microsphere held in optical tweezers. (a) The position of the microsphere was monitored using a four quadrant detector, as the kinesin walked along a fixed microtubule track. (b) Note that the staircase structure to the position data is a direct result of the single kinesin pausing in between individual ATP cycles. For further information, see http://mc11.mcri.ac.uk/motorhome.html.

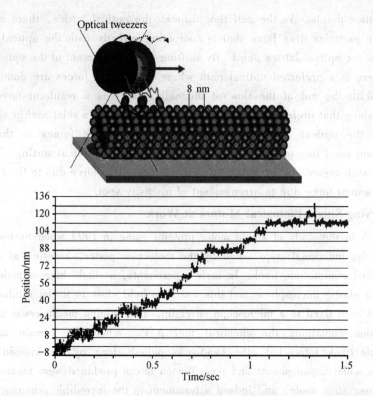

Figure 4 Mechanical recording made from a single processive kinesin, a molecular porter that walks along the microtubule track taking steps that are commensurate with the microtubule 8 nm lattice repeat

3. Viscous Drag Force Calibration

We apply a known force and measure the resulting signal as a function of the applied force. If the position of the bead as a function of force is also measured, the results can then be fit to $f = -kx$ to determine the trap stiffness k.

The force due to viscous drag on a sphere of known radius can be calculated. If a liquid with viscosity η flows past a sphere of radius r with velocity v, the force due to viscous drag F_{vis} is given by Stokes formula:

$$F_{vis} = \gamma v = 6\pi\eta r v$$

where γ is the viscous drag coefficient. Since the viscosity of the liquid is known and polystyrene spheres of known radius can be obtained, we can apply a known force if we can measure the velocity of the liquid. This can be done using a video camera by measuring the force signal at a fixed liquid flow velocity and then suddenly turning off the trap and measuring the rate at which the bead moves along with the liquid. A more precise measurement of the flow velocity can be obtained by trapping a bead in an optical trap without flow and then oscillating the cell holding the liquid that surrounds the trapped bead at a fixed frequency and amplitude. If the position of the cell as a function of time is given by

$$x = x_0 \sin(\omega t)$$

then the velocity of the cell (and therefore the liquid surrounding the bead) is

$$v = \omega x_0 \cos(\omega t)$$

and the force due to viscous drag on the trapped bead is

$$F_{vis} = 6\pi\eta r\omega x_0 \cos(\omega t)$$

In this case all of the quantities on the right side of $F_{vis} = 6\pi\eta\gamma\omega x_0 \cos(\omega t)$ are known, so we know the applied force as a function of time. The measured signal S as a function of time will be

$$S = A\omega\cos(\omega t)$$

Thus, we get

$$F_{vis} = \frac{6\pi\eta r x_0}{A} S \equiv D \cdot S$$

Whenever a signal S is detected with a bead in the trap, we can directly calculate the force on the bead.

[Procedure]

Put a little dry yeast into the narrow neck glass bottle. Dissolve the yeast with water. Swirl the contents until all the solid has dissolved. Transfer appropriate amount of solution to the sample cell in drops. Drop some matching oil onto 100 times objective lens. Place the sample cell into the stage stand.

(1) Plug in the power plugs of the equipments.
(2) Turn on the computer, enter the working state of the video program.
(3) Transfer the microscope illumination light voltage to the lowest. Turn on the switch and adjust to appropriate brightness. Show clear images under the microscope.
(4) Adjust the laser operating current to minimum. Then turn on the power switch and

transfer the current to the appropriate value. Weaken microscope illumination light, fine-tuning the objective lens, find the laser spot on the display. Restore the microscope illumination. *Note: laser operating current should not be too high to breakdown the laser.*

(5) Manipulate: Set a particle close to the optical trap by moving the sample stage. Move the control platform gently to capture the particle and then slowly accelerate it until the particle is out of the optical trap.

(6) Take a video of step (5) by CCD-acquisition card.

(7) Image analysis: open video taken in step (6) and calculate the maximum optical trap force. Taken two images before and after the trapped cell escape, measure the displacement of one fixed point of the sample cell (which can be the stain or sticky cell on the cell bottom). Escape speed can be calculated by the measured displacement and the time interval of two inter-frames. Then calculate the maximum optical trapping force with stokes formula.

(8) Repeat Step (5) ~ (7) for multiple measurements.

[Questions]

1. What's the principle to trap the particles by light and how to achieve that via experiment?

2. List the elements affect the light trap.

3. List the application fields according to the characters of the light trap.

4. If two sphere particles are trapped simultaneously, what is the most likely form to be arranged? Why?

[Appendix]

It is helpful to consider the sensitivity and range of a transducer required to study different biological processes, in terms of the energy, force, length and time.

Energy:

1 photon = 400 pN · nm;
Hydrolysis of 1 adenosine triphosphate (ATP) molecule = 100 pN · nm;
1 ion moving across a biological membrane = 30 pN · nm;
thermal energy = 4 pN · nm.

Force:

Required to rupture a covalent bond = 1 nN;
Required to convert DNA from a double helix to a ladder = 50 pN;
Required to break most protein-protein interactions = 20 pN;
Produced by most motor proteins = 5 pN.

Length:

diameter of a bacterium and optimal size for beads held in optical tweezers = 1 μm;
resolution of light microscope = 300 nm;
diameter of eukaryotic cell organelles = 100 nm;
large protein assemblies and virus particles = 25 nm;
work stroke produced by motor protein = 5 nm;
diameter of hydrogen atom = 0.1 nm.

Time:

cell division = minutes;
cycle time of many biochemical processes = seconds to milliseconds;
individual biochemical steps = milliseconds to microseconds;
protein conformational changes = nanoseconds;
molecular dynamics = picoseconds.

[Appendix]

It is helpful to consider the sensitivity and range of a transducer required to study different biological processes. In terms of the energy, force, length and time:

Energy:
1 photon ≈ 100 pN·nm;
Hydrolysis of 1 adenosine triphosphate (ATP) molecule ≈ 100 pN·nm;
1 ion moving across a biological membrane ≈ 50 pN·nm;
thermal energy ≈ 4 pN·nm.

Force:
Required to rupture a covalent bond ≈ 1 nN;
Required to convert DNA from a double helix to a ladder ≈ 50 pN;
Required to break most protein-protein interactions ≈ 20 pN;
Produced by most motor-proteins ≈ 5 pN.

Length:
diameter of a bacterium and optimal size for bead held in optical tweezers ≈ 1 μm;
resolution of light microscope ≈ 300 nm;
diameter of eukaryotic cell organelles ≈ 100 nm;
large protein assemblies and virus particles ≈ 25 nm;
work stroke produced by motor protein ≈ 5 nm;
diameter of hydrogen atom ≈ 0.1 nm.

Time:
cell division ≈ minutes;
cycle time of many biochemical processes = seconds to milliseconds;
individual biochemical steps = milliseconds to microseconds;
protein conformational changes = nanoseconds;
molecular dynamics = picoseconds.

4.2 Scanning Tunneling Microscope (STM)

In 1986 Gerd Binnig and Heinrich Rohrer were awarded the Nobel Prize for Physics for the groundbreaking invention of the scanning tunneling microscopes (STMs). It generates a highly-resolved image of the specimen.

【Objective】

We will review six important topics in this lab to understand how STM works, which are of great general utility and interest for experimental science:

- How quantum mechanical tunneling works in STM?
- How to control very small displacements using piezoelectric transducers?
- How to use feedback to control tunneling currents?
- How to vibrationally isolate sensitive systems?
- How to collect data electronically?
- How to image process STM data to extract useful information?

【Equipment】

(1) The Nanosurf Easy Scan STM.
(2) An electronics controller box that measures and sets tunneling current, sets the bias voltage, sets the piezo voltages and provides the feedback.
(3) A computer that communicates with the electronic control box to run experiments, assemble STM images, and allow analysis and storage of the images.
(4) A vibration isolation chamber – in this case, a granite block with shock-absorbing feet.

【Principle】

We first consider the case of a massive particle such as an electron which travels along a one-dimensional path from a region with potential energy $V = 0$ to one with potential energy $V = V_0$ = constant (this is known as a step potential). See Figure 1(a). The electron wave is totally reflected from the interface, yet unlike a classical particle, the electron has a finite probability of being found in the classically forbidden region where $E < V_0$. This is because its wavefunction decays to zero exponentially over a distance determined by V_0 and E (Figure 1(b)).

Now, consider the case where the potential energy only equals V_0 over a distance a, after which it drops back down to $V = 0$. This case, is known as a barrier potential. Now, the wavefunction will not in general have decayed to zero when it reaches the other side

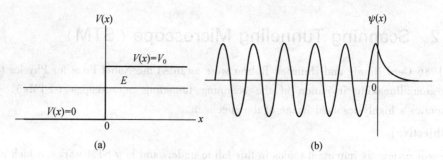

Figure 1 Potential energy functions for a step potential (a) and corresponding wavefunction (b)

of the potential energy barrier. The net result is that the electron wavefunction has nonzero amplitude with probability amplitude T (for transmission) on the other side of the barrier, with approximate dependence:

$$T \cong \exp(-2ka) \tag{1}$$

where $1/k$ is a measure of the distance over which the exponentially varying wavefunction decays within the barrier. It is determined by the values of the particles total energy, E, and the potential energy, $V(x)$, within the barrier by:

$$k = \sqrt{\frac{2m}{\hbar^2}(V_0 - E)} \tag{2}$$

This means that if the electron wavefunction describes a situation in which an electron is incident from the left, it has a probability of *either* being reflected from the barrier or being transmitted, even though it must pass through a classically forbidden region to do so. It is as though a tennis ball thrown against your dorm room wall suddenly disappears from your room and reappears on the other side!

The electrons within an electrical conductor are in states well described by a free particle wave function. As a result, when two conductors are brought very close together yet still separated by an insulating barrier (such as an air gap or layer of insulating oxide), electrons can still flow between them by tunneling. If an electrical circuit is completed between the two conductors, this flow of electrons can be sustained and measured as an electrical current. Just as the transmission coefficient, T, has an exponential dependence on distance, so does the tunneling current depend exponentially upon separation between the two conductors. This is the situation in many common lab settings. If you join two pieces of wires by twisting them together or by sticking them into a breadboard, you often are relying on efficient tunneling across the small gap between them to complete your circuit. This is because you often have thin layers of insulating metal oxides coating the surfaces of copper wires.

This is also what happens in STM. There, one conductor is the very sharp tip of a metal such as tungsten or platinum (with a small 10% admixture of iridium to improve its stiffness). These materials are chosen because you can use them to produce STM tips that have very sharp protrusions ending in only one or a few atoms. Imagine that you get a tip in which one atoms protrudes beyond the others by a few Angstroms, as shown in Figure 2. A sharp tip located on a flexible cantilever is used to probe the distance between the tip and sample surface, as judged by the tunneling current. Since the tunneling current also depends on the chemical nature of sample and tip, the STM also severs for characterization of electronic properties of solid samples.

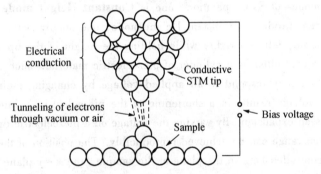

Figure 2 When the distance between conductive tip and conductive sample is lowered to a few Å, electrons can traverse the gap with some transmission probability. The STM measures not purely distance like the AFM, but the local density of electronic states

[Principle of STM Operation]

STM's principle of operation (Figure 3) requires electrically conductive samples. Often the tip radius of the probe is only a few nm or less. A finely sharpened electrically conductive tip is fist positioned within about 1 nm of the sample by mechanical translations stages and the piezoelectric scanner. At this small separation, electrons tunnel though the gap between tip and sample (Figure 2). The tunneling current depends on the applied bias voltage between tip and sample, the distance, the tip shape, and the chemical compositions of sample and tip. The feedback loop ensures constant height or constant current. Tunneling current and feedback voltage are a measure of surface morphology and composition.

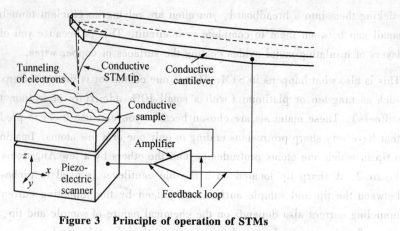

Figure 3 Principle of operation of STMs

There are two modes of STM operation, one is **Constant Height mode**, the other is **Constant Current mode.** In Constant Height mode, one simply scans the tip in the plane of the sample, left and right, while holding the height of the tip constant. The tip's motions are controlled by a cylinder of piezoelectric material. Such materials have the properties that they respond to an applied voltage by changing their dimensions. One polarity of voltage results in a shortening of the piezoelectric, while the opposite polarity induces an expansion. By varying the voltage on a piezoelectric displacement at the sub-Angstrom range can be achieved reproducibly. The position of the STM tip can thus be finely controlled both in the plane of the sample (the $x - y$ plane) and in the z direction.

The tip-sample spacing varies as the tip is scanned horizontally over the sample, because the surface has atomic-level peaks and valleys due to its atomic structure, and so the tunneling current also varies. So, if one measures the tunneling current I_t as a function of in-plane location (x,y), one can "map out" the topography with atomic precision; high current corresponds to a raised area of the sample. The act of repeatedly scanning x and y back-and-forth to yield an image is also called **rastering**. This yields a map of (x,y,I_t) from which an image of the sample's surface can be made.

Since one cannot plot in three dimensions, two methods are used to create such surface plots. Either colors (or shades of gray) are used to indicate current (a scale bar is conventionally printed by the side of the image to indicate correspondences between currents and shadings) or a computer reconstruction of the surface is generated and the image is viewed from an angle to indicate its 3D structure, see Figure 4.

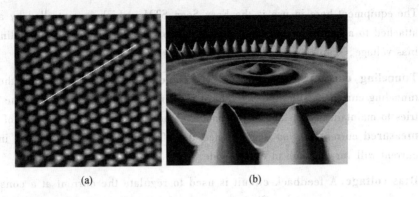

(a)   (b)

Figure 4 Different methods of represent STM measurements. (a) Burleigh instruments image of the surface of graphite, in which gray-scale (shades of gray) are used to indicate height within the plane; (b) IBM image of a "quantum corral" (ring of atoms binding a surface electron) in which computer 3D reconstructions are used to indicate surface structure

From http://researcher.watson.ibm.com.

If you try to use constant height mode to image structures with bumps large compared to the tip-sample spacing, you will hit your tip on the sample and damage it! You can avoid this problem by imaging instead in **Constant Current Mode.** In this mode, feedback is used to fix the current at a constant target value (called the Reference Current) as the piezoelectric is used to scan the tip back-and-forth in-plane. If the sample surface is higher at one point than another, the tunneling current goes up. The feedback circuit responds by retracting the tip so the tunneling current is restored to the reference value. If the sample surface is lower at one point, the tunneling current decreases, and the feedback circuit and uses the piezos to lower the tip until the tunneling current returns to the reference value. In this mode, the *position of the tip*, rather than tunneling current, is recorded to yield the map of the surfaces' topography as $(x, y, \text{tip } z)$. This interplay between measurement (of tunneling current) and regulated control (of tip height, which regulates the tunneling current) is an instance of negative feedback. It's **negative** feedback because a greater tunneling current results in a tip displacement that **reduces** the tunneling current. In other words, the motion of the tip always opposes the change measured in the tunneling current.

As a practical matter, you will begin imaging by using Constant Current mode, then—if you wish to see atomic scale details—switch over to Constant Height mode after you have found a flat region to safely scan this way.

The equipment here in use is the Easy Scan STM, which is controlled by a computer attached to a controller box that allows you to set and monitor the tunneling current, bias voltage, scan range and feedback controls.

Tunneling current. Nanosurf suggests standard working values for the *reference* tunneling current, called the SetPoint of 1.00 nA. This is the average value the system tries to maintain in constant current mode. You will see a working display of the **actual measured** current as you scan. Since the STM must scan in order to image, this current will vary in time in either mode.

Bias voltage. A feedback circuit is used to regulate the current at a constant value in constant current mode. Their suggestion for Gap Voltage (the tip-sample voltage, also called the Bias Voltage) is 0.05 Volts, corresponding to the tip-to-sample voltage.

Scan range. The X-Range setting determines how large the sample area scanned is. You will use a point and click technique to determine *where* on the sample you scan. To start your scans, it is helpful to click "Full" on the top menu. This starts the scans at the largest scan range, 560 nm. You can then zoom in. The Z-Range setting determines over how wide a range of z-values are recorded. This is necessary because the height values are stored as an 8-bit number. An 8-bit number can only represent at most 2^8 values = 256 separate distinct z-values. When you use a large Z-Range, you can measure a wider range of z-values, but your smallest distinguishable step in z is large. A small Z-Range allows finer measurements of z, but your tip can go off-scale more easily by measuring z-values too high or too low to fit into the measurement range.

Feedback control. Your STM's feedback circuit uses a common design known as PID. (for proportional/integral/differential.)

The proportional setting (P-Gain, default value 13) is useful if you are trying to correct for currents far from the set current value. The correction voltage applied to the piezos to correct the current is equal to a gain factor (set by the P-Gain in software) multiplied by ΔI = (actual current)-(reference current). If ΔI is large, the correction voltage is large, and the sign of the correction voltage varies with the sign of ΔI. However, just proportional gain by itself becomes less effective as the current approaches that of its set point, since then $\Delta I \sim 0$ and no information is available to adjust the piezo for good control. In this case, it's useful to move the piezos based on an electronically filtered version of the ΔI signal.

4 INNOVATION EXPERIMENTS

The Integral setting basically has the piezo voltage respond to an integrated version of the ΔI, while the Differential setting changes the piezo voltage based on a differentiated version of ΔI. In filter, terms, the integral mode responds to low frequencies, while the high pass responds to high frequencies in the current signal. The time constant over which the current difference signal is integrated is set by the Integrator I-Gain (default value 13) software control. The net effect is to give information about the time behavior of the way the current difference ΔI changes, to enable the circuit to more effectively maintain the current at its set value. Since a differentiated signal responds strongly to high frequency fluctuations the most, and noise introduces unimportant high frequency fluctuations, the Nanosurf Easy Scan STM uses only integration. The integrator can give a time-integrated feedback to the piezo which helps it track larger-scale outlines of the surface, which helps in tracing out longer range scans with more surface topography.

We have found that the default settings of the proportional and integral gain do well for the gold nanogrid imaging in constant current mode. However, for moving to atomic-scale imaging, you may wish to start with these values, then reduce the I-Gain to around 2 (to preserve some tracking to avoid crashing the tip) and the P-gain to 0 (to avoid having the tip simply move to track the atomic scale features). Also, you may find other values work better for your samples at the atomic-scale; for instance, sometimes it works well to start with the defaults and gradually lower the gains to see what gives the best atomic-scale image quality.

At the submicron scale, it's impossible to avoid a little bit of a tilt even it may not look tilted. This may be so large as to make it hard to see your actual sample's features. You can search the manual how to correct for sample tilt by subtracting out an average slope in the data.

You can see how these work in a good video at the website:
http://www.iap.tuwien.ac.at/www/surface/STM_Gallery/stm_schematic.html

[Procedure]

1. Experiment A: Running the STM in Simulation Mode

(1) Be sure you have the Easy Scan power supply turned OFF, and starting the software in Simulation Mode. This is done automatically if you startup without power to the STM.

(2) Double-click on the Easy Scan E-line icon to start the software, it will say that it cannot find the controller box, and ask if you wish to run in simulation mode.

Answer "Yes", and proceed. Or, when a window appears saying "No connection to microscope!" by clicking on "Start Simulation".

(3) Follow the instruction in the manual and do the exercises. AT NO TIME WILL YOU BE TOUCHING THE STM DURING THESE EXERCISES! Use them to familiarize yourself with the software.

2. Experiment B: The Gold-coated Nanogrid

Find the nanogrid sample and the sample holder. The gold nanogrid is in a clearly marked plastic box that says "Nanogrid" on the lid. You must at no time touch the sample or trip, or sample holder with your hands! If you do, ask your instructor to clean the contaminated surfaces with a clean cotton swab and rubbing alcohol. Take a careful look at your sample before you place it inside your STM. The sample does not cover the entire magnetic mount; instead a dot of silver paint is placed on the side of the sample to provide electrical contact with the magnetic mount. This is because the nanogrid itself (beneath the gold) is not conducting. Without the gold coating and the silver paint, you would get no tunneling current.

Now you are ready to place the sample carefully back into the STM. Follow the manual's instructions mounting samples. Be especially sure not to damage the tip! Be sure to position it so the very center of the gold surface (not the edges) is centered under the tip. Your tip and sample are now a considerable distance apart. Use the magnifying loupe to watch the tip and sample carefully under magnification while you gently use the black knob on the sample holder to move the sample closer to the tip. Decrease this distance to about 1 mm. After you are done, gently place the clear plastic cover back on. You can use its built-in magnifier to check your tip-sample distance again.

Now, follow the manual's instructions and following pages to establish a tunneling current. First, use the Approach Panel manual approach to close up the distance between the tip and the sample to around 0.5 mm or less if you feel comfortable. (This entails holding down the left mouse button while clicking on the down arrow, WHILE CLOSELY WATCHING THE TIP APPROACH THE SAMPLE UNDER THE MICROSCOPE! BE CAREFUL NOT TO CRASH THE TIP) Then, use the automatic approach, as explained in the Easy Scan manual. If you cannot get the STM to tunnel and the red LED stays lit, ask your instructor for help!

Once you have a tunneling current, follow the manual's instructions for imaging the sample. Record the following for your lab report. Indicate what current, voltage and gain settings you used to perform your scans.

(1) Use the "Full" setting in the Scan Panel to image at the largest x and y-range. Correct for sample tilt if you need to. Print out your image.

(2) Then, reduce the Scan Range to zoom in at different magnifications. Do the features on your nanogrid change size the way they should?

(3) Using the largest range, use the Tools menu to measure the spacing of the grid. See the Software manual. Try out at least two different ways to measure the spacing, using the Measure Distance command and the Create a Cross Section command.

(4) Use the instructions on "Achieving Atomic Resolution" on manual to see if you can image gold atoms on the flat tops of the grid. Print out your highest magnification image.

3. Experiment C: HOPG (highly oriented pyrolytic graphite) sample

(1) Image Acquisition: The following low and high-resolution images should be collected in this experiment and included with your laboratory report:

1) 10 mm × 10 mm low-resolution image.
2) 5 mm × 5 mm low-resolution image.
3) 1 mm × 1 mm low-resolution image.
4) 100 nm × 100 nm high-resolution image.
5) 25 nm × 25 nm high-resolution image.
6) 10 nm × 10 nm high-resolution image.
7) A high-resolution image of your choice to show the best atomic-level resolution possible.

Notes:

1) Each image should be optimized prior to image capture, as described in the STM Operating Instructions.

2) If you encounter difficulty achieving atomic resolution, then try changing the X and Y offsets to scan another region. If you are still not successful, then try minimizing the gain values and increasing the scan rate. If neither of the above adjustments result in atomic resolution, then consult your instructor.

(2) Image Processing: For the first three low-resolution images, it may be necessary to use the "Flatten" function available in the "Modify" pull-down menu to correct for any microscopic sample tilt, which is evidenced by a continuous change in contrast across the image in one direction.

However, terraces and step edges may be more clearly defined in height profiles if the "Flatten" function is not used.

For the last four high-resolution images, it may be necessary to use the "Spectrum 2D" fast Fourier transform function, as described in the STM Operating Instructions. Consult your instructor if additional image processing is necessary.

(3) Image Analysis:
1) Measure the average surface roughness of each image using the "roughness" function (rms value) available under the "Analysis" pull-down menu. Prepare a table to summarize your data as well as a graph of the surface roughness as a function of scan size.
2) Use the "Section" function available under the "Analysis" pull-down menu to measure horizontal and vertical distances for your images. For the best low-resolution image, draw a line across the image to plot the depth profile along that line. Measure the length of any horizontal features (terraces) as well as the height of any vertical features (steps). Summarize your results in a table.
3) For the best high-resolution image, draw a line through a row of the primary features (atoms) in the image to plot the depth profile along that line. Measure the horizontal distances for five different adjacent carbon atoms in the image to determine the average distance between the carbon atoms. Also measure the height of five different carbon atoms in the image to determine the average carbon-atom height. Summarize your results in a table.
4) For the best high-resolution image, measure all the angles defined by a particular carbon atom and two of its nearest neighbors. To measures angles, access the "Top View" function available under the "View" pull-down menu. On the second monitor, click the "Angle" function. Using the left mouse button (left click), click a point on the image to represent the center of the angle. Left click a second position with the mouse to define one side of the angle. Left click a third point to define the second side of the angle. The measured angle will appear on a white strip at the bottom of the monitor. Click the right mouse button to terminate the angle-measuring function. Repeat the measurement two more times using different carbon atoms to determine the average angles. Summarize your results in a table.

(4) Image Interpretation:
1) (a) What is the origin of the surface roughness for the best low-resolution image?

(b) If the distance between adjacent graphene sheets is 3.35 Å, how many grapheme layers have been removed for each of the step heights measured, if any, for this image?

2) (a) What is the origin of the surface roughness for the best high-resolution image?

(b) Discuss the plot of surface roughness versus scan size in both the low and high resolution regions.

3) (a) Is the symmetry of the atomic arrangements in the best high-resolution image the same as that expected for the hexagonal rings in graphite?

(b) How do they differ?

4) (a) If the distance between nearest-neighbor carbon atoms in a hexagonal graphene ring is 1.42 Å, construct the graphene ring and superimpose it on the best high resolution image. Are the carbon-atom positions coincident?

(b) Which carbon atoms in the hexagonal graphene ring appear in the best high resolution image?

(c) On the basis of the three-dimensional structure of graphite, try to explain your answers to (a) and (b).

5) (a) Based in your average carbon-carbon distance and angles and assuming that the C-C distance in a hexagonal graphene sheet is 1.42 Å, draw a figure showing the positions of the carbon atoms imaged in this experiment as open circles.

(b) Show the positions of any carbon atoms not imaged in this experiment as shaded circles. How would you propose to image these carbon atoms?

6) Compare your average carbon-atom height with the radius of the carbon atom in the graphene rings. What are the implications of this comparison?

Note: The INVSEE Allotropes of Carbon Module may be helpful to interpret your images 6.

[Questions]

1. What are the basic operating principles of a scanning tunneling microscope operating in
   (1) the constant-current mode and
   (2) the constant-height mode?
   (3) Identify one advantage and one disadvantage of each mode of operation.

2. Draw a basic functional diagram of a scanning tunneling microscope operating in the constant-current mode and explain how an image is obtained.

3. (1) Compare the structures, bonding, and electrical and mechanical properties of graphite and diamond.
   (2) Could STM be used to image both materials? Why?

4. Answer the following questions from the image provided in Figure 2:
   (1) What is the symmetry of the nearest-neighbor carbon atoms surrounding a particular carbon atom?
   (2) What are all the angles defined by a particular carbon atom and two of its nearest neighbors?
   (3) What is the distance (in Å) between two adjacent carbon atoms? Is this distance in good agreement with that shown in Figure 1?

4.3 Measuring Planck Constant

【Objective】

- To verify the photoelectric effect.
- To measure the kinetic energy of the electrons as function of the frequency of the light.
- To determine Plank constant h.
- To show that the kinetic energy of the electrons is independent of the intensity of the light.

【Principle】

Electron can be liberated from the surface of certain metals by irradiating them with light of a sufficiently short wavelength (photoelectric effect). Their energy depends on the frequency ν of the incident light, but not on the intensity; the intensity only determines the number of liberated electrons (Figure 1). This fact contradicts the principles of the classical physics, and was first interpreted in 1905 by Albert Einstein. He postulated that light consists of a flux of particles, called photons, whose energy E is proportional to the frequency f:

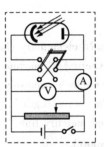

Figure 1 Schematic representation for measuring Planck's constant h using photoelectric effect. Monochromatic light falls on cathode of a photocell

$$E = hf \tag{1}$$

The proportionality factor h is known as Planck's constant and is regarded as a constant of nature. Each photoelectron is replaced by a photon and exists the atom with the kinetic energy

$$E_{kin} = hf - W_K \tag{2}$$

Where W_K is the work function of the electrons. It is independent of the irradiated material.

Some of the ejected photoelectrons travel to the anode, where they are registered in the form of a photoelectric current I. If the photoelectrons are ejected against a negative potential which is gradually increased, the photoelectric current continually decreases. The voltage at which the photoelectric current reaches precisely zero is called the limit *voltage* U_0. It can be used to calculate the kinetic energy of these weakly bound electrons:

$$eU_0 = hf - W \tag{3}$$

where e is the elementary charge.

The measurements are conducted for various wavelengths λ and frequencies f:

$$f = c/\lambda \tag{4}$$

where c is the speed of light in vacuum. When we plot the limit voltage $U_0(f)$ as a function of f, we get a straight line with the slope:

$$\Delta U_0 / \Delta f = h/c \tag{5}$$

For a known elementary charge e, this gives Planck's constant h.

In this experiment, narrow-band interference filters are used to select the wavelengths, each filter selects precisely one spectral line from the light of a high-pressure mercury lamp. The wavelength specification on the filter refers to the wavelength of the transmitted mercury line.

[Equipment]

Device with photo cell for determining Planck's constant, high mercury lamp with socket, universal choke in housing, lens in holder, iris diaphragm in holder, filter revolver, interference filters of 405 nm, 435 nm, 546 nm and 578 nm, electrometer amplifier, plug-in unit, 100 pF capacitor, switch, DC voltmeter, optical bench with riders, various plugs, wire leads, outlet box.

[Procedure]

Note: The high pressure mercury lamp also emits light in the ultraviolet range, and can damage the eyes! Never look into the direct or reflected light from the mercury lamp! The high-pressure mercury lamp reached its full intensity after a ten-minute warm-up period.

You do not need to darken the room; this has no effect on the measurement results.

Set up the various components as shown in Figure 2. Their separations are given in cm.

(1) Mount the high pressure mercury lamp a at the left edge of the track and connect it to the universal choke, plug the choke into the outlet box and the outlet box into the floor outlet and switch everything on. This is so that you can start measuring as soon as you are finished.

(2) Mount the photocell e at the marked position using a 90 mm rider; remove the cover align the photocell so that the coated black surface is facing the mercury lamp. Mount the iris diaphragm b on the optical bench at the marked position using a 120 mm optical rider. Mount the lens c at the marked position using a 120 mm rider and adjust its height so that the center of the lens is at the same height as the center of the iris diaphragm. The light from the mercury lamp should now produce a sharp light spot on the black coating, the sensitive area, of the photocell. The light should NOT fall on the metal ring nor on the part of the black-coated area to which the contacts are attached. The edge zones should not in illuminated either.

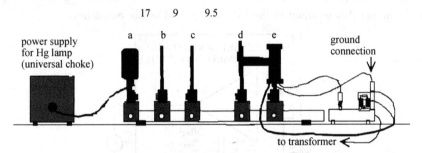

Figure 2 Set up of the measurement

To ensure that this is so, carry out the following procedure, repeating as often as necessary to produce the optimum image:

Vary the height of the iris diaphragm and the lens so that the light spot falls on the black zone of the photo cell, make sure that the center of the lens is always on the same level with that of the iris diaphragm. You may also need to adjust the height and inclination of the photocell (using the screws below the base).

Using the iris diaphragm, adjust the size of the light spot so that it illuminated the largest possible area of the black zone of the photocell, without shining on the outer zones, the metal ring or the contacts on the black coating.

Focus the light spot as necessary by moving the lens along the optical bench.

Note: Once you have adjusted the experiment setup, be sure not to change the setup again.

(3) Place the cover on the photocell.

Place the filter revolver d with iris diaphragm directly in front of the photocell using a 120 mm rider and connect the iris diaphragm of the filter revolver with the cover of the photo cell to prevent scattered light from reaching the photocell.

(4) Set up the electrometer amplifier circuit as shown in Figure 3. As f is the terminal plug, the 100 pF capacitor and the key switch. At g is the coupling plug, the BNC/C mm adapter and the straight BNC cable, which connects to the gray shielded cable of the photocell. At h is connected both black cables of the photocell and the wire to the base of the photocell. At i and j is the multimeter, set to read voltage. Connect the plug-in supply unit to the electrometer amplifier and plug it in via the outlet box. Connect the optical bench (and possibly the rod of the basic device of the photocell) to the ground connection h of the electrometer amplifier and connect this terminal to the external ground of the outlet box.

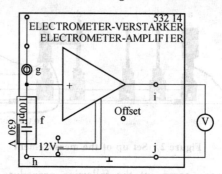

Figure 3 Electrometer amplifier circuit

(5) Switch on the multimeter and set the range switch to 1 VDC.
(6) Turn the interference filter for yellow light (λ = 578 nm) into the beam path.
(7) Discharge the capacitor by holding down the key switch until the multimeter reads zero volts.
(8) Start the measurement by releasing the key switch; wait about 30 s to 2 minutes until the capacitor has charged to the limit voltage U_0. Write down the measured value for U_0. Record the reading in Table1.
(9) Turn the interference filter for green light (λ = 546 nm) into the beam path and repeat the measurement. Extend the measuring range to 3 V and repeat the measurement with the blue (λ = 436 nm) and violet (λ = 405 nm) interference filters.
(10) Vary the intensity of the incident light at the photocell using the iris diaphragm of the filter revolver and measure and record the limit voltage U_0 for each setting.

Note: If the iris diaphragm is closed too far, this may affect the uniform illumination of the light spot on the cathode. Also, leakage currents will play an increasing role.

[Data and Analysis]

Table 1 The limit voltage and frequency for different lights

Colour	$\lambda/10^{-9}$ m	$f/10^{12}$ Hz	$U_0/$V
Yellow	578		
Green	546		
Blue	436		
Violet	405		

Plot a graph of U_0 vs f and find the slope according to Eq(5).

The slope of plot $\Delta U_0/\Delta f =$ _____ .

Plank constant is determined as $h =$ _____ .

[Data and Analysis]

Table 1. The Hall voltage and frequency for different lights

Colour	×10¹⁴ Hz	V
Yellow	575	
Green	540	
Blue	430	
Violet	60?	

Plot a graph of ΔV vs f and find the slope according to Eq.(5).

The slope = plot ΔV vs f = _____

Planck constant is determined as h = _____

APPENDIX

Vocabulary

absolute error 绝对误差
accuracy 准确度
adiabatic expansion 绝热膨胀
air track 气垫导轨
ammeters 安培表
amplitude 振幅
angular frequency 角频率
antinodes 波腹
attenuation 衰减
average(mean) deviation 平均偏差
capacitance 电容
cathode ray oscilloscope 阴极射线示波器
constructively interfere 相加干涉
curve fitting 曲线拟合
destructively interfere 相消干涉
determinate (or systematic) errors 系统误差
deviation 偏差
diffraction grating 衍射光栅
diffraction 衍射
diode 二极管
direct current 直流电
discrepancy 差值、不同
elastic collision 弹性碰撞
elastic modulus 弹性模量
electromotive force 电动势
enantiomer 对映体

error analysis 误差分析
error propagation 误差传递
error 误差
frequency 频率
fundamental 基波、基频
geometric slope 几何斜率
grating constant 光栅常数
harmonic analysis 谐波分析
harmonic series 泛音列、调和级数
harmonics 谐波
heat capacity 热容
impedance 阻抗
in phase 同相
indeterminate (or random) errors 随机误差
isentropic process 等熵过程
isobaric process 等压过程
isochoric process 等容过程
isothermal process 等温过程
limits of error (or maximum error) 最大误差
longitudinal waves 纵波
micrometer screw gauge 螺旋测微器
millisecond meter 毫秒计时器
momentum conservation 动量守恒
monochromatic coherent light 单色相干光
multimeter 万用表

nodes 波节
Ohm's law 欧姆定律
out of phase 反相
overtones 泛音
period 周期
phase angle 相位角
potential（voltage）电压
power（wattage）功率
precision 精度
reactance 电抗
relative（or fractional）error 相对误差
resistance 电阻
resonance 共振
reversible adiabatic process 可逆绝热过程
rigid body 刚体
rotational inertia 转动惯量
significant figures 有效数字
spectrum 光谱
standard deviation of the mean 平均标准偏差
standing wave 驻波

strain 张力
stress 压力
the method of least squares linear regression 最小二乘法
the standard deviation（root-mean square）标准均方根
the superposition principle 叠加原理
traction poise 牵引砝码
transformer 变压器
transverse wave 横波
true value 真值
uncertainty 不确定度
vernier caliper 千分尺
voltmeters 伏特表
voltmeter 电压表
wavelength 波长
the method of weighted least squares 加权最小二乘法
weighted successive differences 连续均差
Young's modulus 杨氏模量

REFERENCES

1. "Conceptual Physics" (10th Edition), Paul G. Hewitt, Pearson Education, 2005
2. "Introduction to Physics in Modern Medicine" (2nd Edition) Suzanne Amador Kane, CRC Press (Taylor & Francis Group), Boca Raton, FL, May 1, 2009
3. "Methods in modern biophysics" (3rd Edition) Bengt Nolting, Springer-Verlage Berlin Heidelberg, 2009
4. Clemson University, "Physics Instructional Laboratories", 2013
5. Georgia Perimeter College, "Principles of Physics I Laboratory", 2008
6. Haverford University medical physics course, 2009
7. Michigan State University, "General Physics Laboratory Manual", 2008
8. Penn State University, "Introductory Physics II", 2013
9. ST. JOHN's University, "Introduction to Physics", 2008
10. University of Chicago, "Atomic and Molecular Lab.", 2010
11. University of Cincinnati, "Introduction to Physics", 2010
12. Wake Forest University, "General Physics II Laboratory Manual", 2013
13. G. T. Caneba, C. Dutta, V. Agrawal, et al., Journal of Minerals & Materials Characterization & Engineering, Vol. 9, No. 3, pp. 165-181, 2010
14. 王家慧,张连娣,等.大学物理实验教程.3版.北京:机械工业出版社,2010

REFERENCE WEBSITES

http://www.darienps.org
http://www.docstoc.com
http://en.wikipedia.org
http://www.fmgvacpump.com
http://www.fofweb.com
http://www.hantek.com
http://www.hotfrog.com
http://www.lambdasys.com
http://www.ld-didactic.de/
http://www.lepla.org
http://www.lhup.edu/~dsimanek/
http://www.loretobalbriggan.ie
http://www.pbf.unizg.hr/
http://physics.about.com
http://www.sjtu.edu.cn